Water Management: Multiple Dimensions

Water Management

Multiple Dimensions

Edited by

RAKESH HOOJA
RAMESH K. ARORA
K.K. PARNAMI

RAWAT PUBLICATIONS

Jaipur • New Delhi • Bangalore • Mumbai • Hyderabad • Guwahati

in association with
Association of State Training Institutions in India

ISBN 81-316-0151-X

Published by
Prem Rawat for *Rawat Publications*
Satyam Apts., Sector 3, Jawahar Nagar, Jaipur 302 004 (India)
Phone: 0141 265 1748 / 7006 Fax: 0141 265 1748
E-mail: info@rawatbooks.com
Website: www.rawatbooks.com

Delhi Office
4858/24, Ansari Road, Daryaganj, New Delhi 110 002
Phone: 011-23263290

Also at Bangalore, Mumbai, Hyderabad and Guwahati

Typeset by Rawat Computers, Jaipur
Printed at Chaman Enterprises, New Delhi

Contents

Acknowledgements

WITH increasing populations and the increased pressure on, and competition for, water for varying uses and for different regions, and with much of the water-related infrastructure developed throughout the 19th and 20th centuries requiring major renovation and modernization or significant maintenance and repair, the optimal use of water has become extremely vital. Major technical, administrative, managerial and social interventions in the water sector along with significant but relevant investment of funds are called for as more and more areas become water stressed.

We, at Harish Chandra Mathur Rajasthan State Institute of Public Administration (HCM RIPA), have long been publishing a journal *Prashasnika*, and since the Association of State Training Institutions in India (ASTI) has its headquarters at HCM RIPA, we also now bring out the journal *ASTI Journal of Training and Development* on behalf of ASTI which is a registered society. We often received articles on various aspects of water and water management for the two journals. For the January June 2005 issues of *ASTI Journal of Training and Development* the three editors of this book could collect enough papers to devote the full issue to water management. Some more papers received for *Prashasnika* or authored by the faculty and friends of HCM RIPA and ASTI led us to feel that we had sufficient material to put together a useful and meaningful book on the multiple dimensions of water management in India in its various facets. This volume is the result.

I am grateful to my two co-editors who along with me have been editing the two journals mentioned above for over two years since June 2005. In fact, the association of Dr Ramesh K. Arora and Mr K.K. Parnami with the editing of the two journals goes back many years during which period I was only a contributor of articles and reviews. But I can more than match them in terms of longevity and intensity when it comes to my passion for reading about, trying to understanding, and writing about water-related issues, particularly in the context of rural India.

Prof. Arora and Mr Parnami join me in expressing gratitude to the countless authors who have over the years sent in their contributions for our two journals. We are particularly grateful to those papers figure in this book.

Also to be thanked is Shri Pranit Rawat, our publisher, who showed great keenness in bringing out the book once he saw the proposed contents.

— **Rakesh Hooja**

Chapter 1

Introduction

Water Management: Multiple Dimensions

Rakesh Hooja, Ramesh K. Arora and K.K. Parnami

WATER, which the eminent Greek philosopher Pindar called the best of all things, is the elixir of life. Its significance to human existence and growth has made it a subject of profound analysis in both the natural and social sciences. More than one book on water has quoted a 16th century German physician Von Hobenheim, known as Peracelsus to the effect that water has been the matrix of the world and of all its creatures.

The total amount of water on earth (liquid or frozen) is about 1,400 million cubic kilometres, of which 97 per cent is in the salty oceans, which cover about three-fourths of the earth's surface. Paradoxically, fresh water available for human use is less than one per cent of the total water on earth.

The then President of India, Dr K.R. Narayanan, while inaugurating a conference organized by the Centre for Science and Environment on "Potential Water Harvesting" had as far back as October 1998 warned that the world was heading towards a "water shock" that could dwarf the oil crisis. Some 80 countries and 40 per cent of the world population are already experiences in water stress. Environmental degradation and population explosion are making the problem even more acute.

India with about 20 per cent of the world's population has to make do with about 4 per cent of its fresh water resource. Out of India's cultivated area, 40 per cent is irrigated, producing 55 per cent of the country's food, while 60 per cent is rainfed, producing the remaining 45 per cent. Further, within the net irrigated area, 31 per cent is covered by canal water, 58 per cent by groundwater and 11 per cent by tanks and other resources.

However, as the World Water Council's *World Water Vision* (2000) points out, today's water crisis is not about having too little water, but it is a crisis of management. As the *Vision for Integrated Water Resources Development* of the Government of India (2003) points out, few issues have greater impact on economic development than the management of its water resources. Similarly, the World Bank document *Water Resources and Sector Strategy* (2004) emphasizes that water resources management and development is central to sustainable growth and poverty management.

For many years now it has been recognized that water management and development is going to be an increasingly complex and sensitive task in the 21st century involving various facets and dimensions. Thus, the importance of effective management of water cannot be overemphasized. Accordingly, certain critical and crucial issues pertaining to water management in the Indian context as discussed by scholars, researchers, policy makers and implementors have been put together in the pages that follow.

Water Demand: Perspective Planning

Yoginder K. Alagh, in his contribution "Long-term Water Demand Projection for India and Possible Policy Responses: A Note" provides a perspective for water demands in the near future. His projections are based on various empirical researches undertaken by him and a few eminent institutions. It is projected that by the end of the present decade, India would have used up most of its balanced water reserves with the irrigated area reaching around 114 million hectares by 2010. The projections for 2020 are: a requirement of water for 122 million hectares for irrigation. Alagh avers that there is need for a serious debate on the issue since the balance groundwater resources are now limited and therefore a very dramatic effort will be needed to harvest and carefully use the available water. If it is not done, the projected increase in cropping intensity will not take place. It is pointed out that in the next two decades, the cropping intensity can increase from 1.3 to 1.5. Thus, harvesting of rainwater, recycling water from agriculture drainage systems and increasingly judicious use of water for cropping will be required. Further, non-agricultural use of water will have to be more economical.

Alagh has pointed out that there will be increasing demand of water in industry as well as the power sector. Cumulatively, therefore, the scarcity of water is bound to assume enormous proportions. What makes the matter worse is underutilization of surface water capacity due to low water use efficiency and an overextraction of groundwater. Thus, an unbalanced growth becomes the cause of unsustainability of water resources in India. It has also been pointed out that the present levels of urban waste water and affluent treatment are bound to cause a deterioration in the water quality, significantly impacting its availability in certain areas.

Alagh endorses the view that a large number of interventions will be needed in order to achieve sustainable water development in tune with sectoral requirement arising from a 7 to 8 per cent annual growth in GDP. These interventions could be technological, institutional or supply augmenting possibilities. Among the various interventions suggested for facilitating sustainability are watershed management, drip irrigation, water users associations, stress on economic cost and other appropriate measures, keeping in view the special needs of the land. The Commission on Water Perspectives has projected balanced demands/supplies in 2025. These perspectives include improvement in irrigation efficiencies to about 60 per cent from the existing levels of less than 40 per cent, conjunctive water use, aquifer management, demand management and rationalization of cropping patterns, greater priority to resource allocation to the water sector, improved project management and phasing, and watershed management. Further, Alagh subscribes to the view that private investment in groundwater promotion needs greater support. Particularly in the eastern regions of India, basin development projects should be undertaken. In the whole enterprise of water management, the role of the cooperatives, NGOs and non-profit companies must increase. Besides, there should be a leadership group to supervise the preparation and implementation of water management programme. It is hoped that with the improvement of investment processes in developing countries in this regard, international financial institutions would play a greater role in promoting sustainability in the projects that would facilitate the improvement of long-term water availability in India.

Savings and Conservation

S.A. Kulkarni in his essay entitled "Water Savings and Conservation in Irrigated Agriculture" stresses that India needs to adopt the strategy of Integrated Water Resource Development and Management (IWRDM) which stands for the process of promoting the coordinated development and management of water, land and related resources with a view to maximize socio-economic welfare in an equitable manner along with promoting sustainable eco-system in order to meet the increasing water demand from irrigation, domestic and industrial sectors. This strategy gains greater significance keeping in view the likely demand caused by the population growth by 2050. Kulkarni is critical of the present irrigation development that has led to waterlogging, salinity and excessive exploitation of groundwater and all these problems have adversely affected the eco-system. Hence, he rightly suggests that we should be categorically clear about the consumptive, non-consumptive, beneficial and non-beneficial components of water use at project, sub-basin and basin levels. This would help us in identifying areas where water saving and conservation can be effected appropriately. The need is to develop rigorous plans and implement them for achieving the goals of saving and conservation. It would also be desirable to look at the international experiences in this context and transplant those which appear to be appropriate in our specific conditions.

Kulkarni observes that what is needed is a significant improvement in the water availability for irrigation and in water productivity standards. For this purpose, all the institutions engaged in the overall water management system have to become forward-looking and creative. In case we wish farmers to be in a more pro-active mode in adopting technologies for water savings, we have to offer them appropriate incentives. Overall, the water resource development and management strategies should aim at improving the water needs of food, people and eco-system at the level of population.

Kulkarni advocates certain water saving and conservation measures in irrigated agriculture. In the engineering domain, these include modernization of canals and control structures, construction of en-route water storages, laser land levelling, and micro and sprinkler irrigation. In the agronomic domain, these measures include appropriate use of poor quality water, dry seeding, deficit irrigation, use and control of weeds, early maturing crop varieties, augmenting non-water inputs and upgrading of rainfed agriculture by supplemental irrigation. Finally, in the managerial realm, a few measures suggested are: benchmarking of irrigation schemes and water auditing, participatory irrigation management, effective implementation of water laws, actively promoting mass-scale public awareness programmes, involving stakeholders, revamping irrigation extension services, effectively using media for propagating water savings and conservation, and updating of water resource database.

The Rural Scenario: A Multi-Pronged Strategy

In his paper entitled "Managing Water in Rural Areas: An Institutional and Multi-use Perspective", Rakesh Hooja discusses the varying uses of water in the rural areas like irrigation, domestic drinking water for people as well as livestock, rural industries and sundry environmental uses.

Although there is a multitude of institutions engaged in water management in rural areas, many of which have been listed by the author, there is greater need to promote Water User Associations (WUAs) and strengthen the concept of Participatory Irrigation Management (PIM). No doubt, the conventional water management, water supply and related institutions also need to modify their operational styles. The state governments should create a facilitating environment, whereby the government departments reorient their working and develop genuine working relationships and partnerships with the WUAs in canal irrigated areas and with watershed committees in rainfed areas. Further, Hooja advocates that canal irrigated areas, rainfed areas and groundwater areas need to be viewed as one continuum and be planned for in an integrated manner. This would imply the need for intensive coordinational interaction among all the concerned government departments and helping them to obtain relevant planning and coordinational inputs from the community-based organizations, user groups and Panchayati Raj bodies.

At a more operational level, it would be feasible to create an integrated district level water resource committee under the District Collector. Hooja argues that the District Collector, being at the helm of regulatory as well as developmental administration, has more competence and authority to coordinate and guide inter-agency water management within the district. We need a paradigm of water governance that works, particularly at the local and regional levels.

Environmental Concerns

Brij Gopal in his essay entitled "Water Management: An Environmental Perspective" points out that water is different from all other resources, for it is both a renewable as well as a non-renewable resource. It is renewable because it cycles continuously between the ocean, air, land and the organisms. However, it is non-renewable because its quantum cannot be increased. Thus, there is need for placing highest emphasis on the quality aspects of water management. On behalf of the various speechless stakeholders – plants, animals and micro organisms – the author emphasizes the need for making available quality water for all living beings.

The most common water quality problems are: oxygen depletion and consequent changes due to discharge of organic wastes, eutrophication from the abundance of nutrients and the existence of toxic substances including heavy metals, salinization and acidification. Since all streams have only a limited capacity for self-purification, there is always a danger of quality depletion in water. Water quality problems also arise when inputs of wastes exceeds the assimilative capacity of rivers and streams. Water impurities also get extenuated by the irrational action of human beings.

Brij Gopal rightly stresses that the current water management practices that seek to increase the productivity of land and support the livelihood of the farmers do not pay adequate emphasis on the productivity of natural water bodies. Hence, it is essential that systematic efforts be made to prevent degeneration of water and generation of waste water through its efficient utilization and recycling. Further, the consumption of water must be reduced in domestic, agricultural and industrial segments. Appropriate devices would have to be taken recourse to in order to meet these ends. There is now a more urgent need for restoring the natural eco-system and let nature function to treat the wastes without any provision for external subsidies. It is a truism that technologies alone cannot provide solutions to ecological problems. There is dire need for respecting and restoring nature's 'ecological integrity'. Accordingly, water management must treat water bodies as 'eco-systems' interacting with their catchments on a river basin.

Community's Role

G.S. Narwani in his essay "Community Water Management" explains that the National Water Policy of 1987 advocated the participation of farmers in the management of irrigation systems, particularly in the distribution of water and in the collection of water rates. Since water is a social asset, it needs to be managed socially. It is in this perspective that community water management has become a crucial policy plank in a large number of countries. Narwani points out that the major objectives of community water management are:

1. Right of water to the community and responsibility of the community to conserve water and to distribute equitably among members.
2. Create awareness regarding associating people with rivers rather than joining of inter-state rivers.
3. Recharging groundwater through sharing past experiences of benefits of water conservation projects executed by the community.

4. Get such water policy prepared at the government level in which community participation is ensured and water is treated as a joint property with the community.
5. Right to use of water for those who conserve water.
6. Create awareness among people regarding water conservation and set up community forums for water management.

The question arises as to what should then be the role of government vis-à-vis farmers in the sphere of water management. Clearly, the answer lies in the premise that the role of government should evolve from that of being a systems manager to that of a service provider and that the farmers should graduate from being mere water recipients to water users and then as water co-managers of the irrigation department. Pani Panchayat in Maharashtra has exemplified how farmers have made an optimum utilization of irrigation waters and assured irrigation of at least two acres to each farmer in drought conditions. Likewise, the Water Users Associations (WUAs) in various parts of the country have proved effective institutions of water management. *Kuhla* in Himachal Pradesh, tubewells in West Bengal and Uttar Pradesh, and tanks in Tamil Nadu through community management provide successful illustrations of this kind.

The Hanumanthaiya Committee constituted by the Government of India (1994) had observed that community participation in soil and water management would be augmented only when the people perceive direct intangible benefits for themselves and understand the technology involved. Further, the fuller participation of the watershed committees would be facilitated when a programme is built up on their accumulated wisdom, skills and systems.

Narwani further points out the constraints in the successful implementation of minor irrigation through people's participation and in this respect advocates a multi-pronged strategy involving awareness generation and resource mobilization. For this purpose, PRIs are considered as the most effective instruments for taking initiative and ensuring proper follow-up of the programme.

Regarding potable drinking water in rural areas, especially during the summer months, it would require a rational use of groundwater, adequate operational maintenance through people's participation and sustaining traditional water management practices. Thus, a comprehensive approach can help in facilitating access to safe drinking water by all the sections of community. For this purpose, the administrators have to act as facilitators and the NGOs will have to come forward to act as catalysts for empowering the stakeholders to manage community resources.

Dealing with community participation in integrated water resources management, M. V. S. Prasad in his essay suggests a specific policy frame. On the basis of experiences in Asia and Africa, it can be surmised that participation in water management can be used as a rallying point for the community to come together. Some of the essential policy premises in this context are:

- The benefiting target group needs to be clearly defined.
- The financial, technological and the managerial aspects of community-based organizations have to be integrated in a participative programme.
- NGOs can be roped in for information and motivation.
- Community leadership needs constant promotion.
- The leaders and members of the community should be properly trained in undertaking the management processes.

Prasad believes that water management should de-emphasize the role of the state as well as the market and instead, the autonomous local groups should be brought to the centre stage. The principle of checks and balances should permeate various local groups, government actors, NGOs and aid agencies. Greater authority and power should be devolved on the local groups. In fact, these local groups should evolve as federated structures of community user groups, capable of negotiating with the government actors on terms of equality. Adequate representation ought to be given to marginal groups in the decision-making mechanisms employed by community organizations. The importance of ensuring accountability of the various decisional groups cannot be underestimated. Only then the centralizing tendencies can be checked. Importantly, local groups should have access to adequate funds. And, these funds should be raised locally and not treated as subsidy or grants from the Central Government.

Kalyan Paul, in his essay "Community Managed Drinking Water Systems in the Himalayas: An NGO Perspective" observes on the basis of field research on community participation that the communities would be more than able to manage their lives provided they are given opportunity and freedom to do so. This would require villages being drawn into innovative dialogues on ideas on transformation in order to enable them to make desirable choices of technology. This can be facilitated, *inter alia*, through farmers' induction programmes. In the Himalayan region, artisans guilds and barefoot engineers have been able to lend support to the communities in managing the drinking water system. Hence, it is high time that the state governments appreciated and recognized the significant role of community organizations in water management.

The Rajasthan Case

A.S. Kapoor in his essay entitled "Water Scenario in Rajasthan" points out that Rajasthan covers 10.4 per cent of the country's area, it has 5.65 per cent of the nation's population, and yet, it has only 1.11 per cent of surface water and 1.72 per cent of groundwater. The per capita availability of water in Rajasthan is one-fourth of the country's average. No wonder, in Rajasthan, water scarcity is acute and is likely to get worse with the passage of time. Ineffective and half-hearted measures to control population have further aggravated water scarcity in the state. Kapoor rightly suggests that vigorous steps for population control are a prerequisite to overcome poverty and water scarcity in the state.

The National Water Resource Council for the first time adopted the National Water Policy in the year 1987 which was later modified in 2002. The Water Policy of the Government of Rajasthan is drafted on similar lines. Though the national and state water policies cover most aspects of water management, these do not deal with the sources of water scarcity and the means of overcoming it. Besides, there is hardly any analysis in the policies on the issue of prioritization in utilizing water.

In Rajasthan, there can be a substantial saving of water through proper management and adoption of water saving measures like lining of canals and use of sprinkler/drip systems. This can lead to efficiency of water use by 70 per cent. The situation can further improve through proper upkeep and maintenance of irrigation works, involvement and active participation of water users in planning, implementation, operation and maintenance. Further, conjunctive use of water in the command areas of large projects should be provided. Kapoor also recommends recharging of groundwater, water harvesting in the countryside and rooftop rainwater harvesting in urban areas.

Kapoor argues that the handicaps of water scarcity in Rajasthan can be wholly overcome for all times through inter-basin transfer of sufficient quantity of water towards the west. Hence, Sharda-Yamuna link is of crucial significance for Rajasthan. He strongly advocates the interlinking of river projects and favours their expeditious finalization and execution. These projects should specially cater to the specific needs of Rajasthan.

HRD Perspective

An important area of water management is the human resource development of those engaged in this critical endeavour. Rakesh Hooja in his essay "Training for Participatory Irrigation Management in India" reveals that Mexico, the Philippines, Colombia and Turkey are some of the well-known examples of transfer or turnover of irrigation systems through PIM (Participatory Irrigation Management). Conversely, in India, the scale and impact of this strategy has been woefully meagre.

Training for PIM should be aimed at developing a cadre of people committed to PIM, reorienting attitudes and behaviour to suit the new paradigms in PIM and to enable the participants to acquire practical knowledge and skills for various PIM-related purposes. Thus, Hooja's paradigm of training is not confined to the traditional modes of training and instead it adopts a wider view of it.

Hooja advocates a comprehensive PIM training that would include organizing workshops for politicians, government functionaries, office bearers of water users associations and NGOs through a battery of methodologies including appreciation workshops and case studies. The focus of this training would be on knowledge enhancement, skill development and attitudinal change. Moreover, there would be an added emphasis on adopting a creative approach to overall training management.

Hooja believes that initiative and leadership for PIM training have to emanate from the Central Government, for it can take a lead in developing infrastructure, facilitating institutional networking, organizing workshops, designing model training modules and devising strategies for executing guidelines in the realm of PIM training. The state governments are also expected to be active players. Therefore, a multi-agency/multidisciplinary state level PIM Training Committee should be set up in each state. Besides, there is a need for designing state-level PIM training and orientation manuals with adequate attention to various regional requirements. Training would be expected to be imparted through formal as well as informal channels.

Hooja argues that there is need to ensure speedy utilization of PIM in all irrigation systems and therefore PIM training and orientation need to receive urgent attention of policy makers.

Watershed Approach

C. Sekar and S. Senthilnathan in their essay entitled "Watershed Approach: A Viable Proposition for Water Management in Rainfed Areas" underline the importance of global rainfed farming in food production to compensate for the reduced investments in irrigation and decreasing groundwater pumping to eliminate groundwater overdraft. They project that both the rainfed area yield and the irrigated area yield are likely to increase constantly in the next two decades. Interestingly, rainfed cereal area is likely to account for the majority of total harvested area and total cereal production.

In India, rainfed agriculture is complex, diverse and risk-prone and is characterized by low level productivity and low input usage. Hence, rainfed agriculture is likely to face several challenges in the years to come. An important challenge in this respect is to transform rainfed farming into more sustainable and productive systems and to better support the population dependent on it. Rainfed areas in India have considerable production potentials which have not been fully exploited due to restricted availability of water and low soil fertility. Besides, the basic problem of rainfed area is degradation of natural resource base like soil. This leads to an overexploitation of groundwater with a long-term negative impact.

In this context, the need is for the conservation of water and strengthening of watershed activities in rainfed areas. Further, there should be expansion of network of such projects and sound efforts to reduce cost and increase returns and productivity of watershed approaches. Further, research targeted to rainfed areas should be accompanied by an increased investment in rural infrastructure. Its objective should be to bridge the gap between potential yields in rainfed areas and the actual yields achieved by farmers. Further, this would require appropriate conservation mechanisms, access to markets, credit input supplies etc. Success in the watershed approach in rainfed areas would depend upon a strong partnership between agricultural researchers, local organizations, farmers, community leaders, NGOs, policy makers and other supporting agencies.

Rakshat Hooja, while commenting upon an ICRISAT/IWMI report synthesizing the knowledge gained about watershed management in India, refers to a typology of participation in watershed projects. This typology includes passive participation, participation in information giving, participation by consultation, participation for material resources, functional participation, interactive participation and self-mobilization. The last one refers to people themselves taking initiative of participating in watershed projects even without the assistance of any external change agent. Rakshat favours a close interlinkage between watershed development paradigms and national level policies related to land, water, forest development and overall water management programmes.

A New Federalism

D. K. Verma and A. Sohrot, in their essay "Conservation and Management of Water Resources: Issues Concerning Inequalities and Policy Intervention for Sustainable Development" argue that in order to meet the needs of existing and future populations and to ensure that habitats and eco-system are protected, water resources must be conserved for society in general and for weaker sections in particular. Rightly, therefore, a sound resource management philosophy together with careful policy and planning for the use of water are essential for effecting an ecological balance. In this wider perspective, the authors have explicated the views of Dr B. R. Ambedkar on the conservation and sustainable developments of water resources, particularly on interlinking of rivers and multipurpose use of water resources.

Verma and Sohrot not only favour augmentation of water resources but also their greater accessibility in equal measures to all sections of society. In this backdrop of egalitarian ideology of water management, the authors present case studies that respectively deal with drinking water supply in a small town, industrial water use, exploitation of groundwater and the systemic approach to the solution of the drought problem. These case studies underscore the theme that process of sustainable development has to be consistent with justice to nature, justice in respect of socially, economically and ecologically disadvantaged sections of society and finally justice that is conducive to intra- as well as inter-generational health and happiness. Hence, for meeting the needs of both the present and the future generations, a substantial shift in values, attitudes and behaviour should be planned.

It may be recalled that Dr Ambedkar, while considering the issues of availability and accessibility of water resources, favoured a more effective and controlling role of the Central Government as against of the state governments, over these resources. Significantly enough, he subscribed to the view that inter-state waterways should be declared as national resources. Accordingly, Verma and Sohrot favour the crystallization of a National Policy for Water to be designed and executed by the Central Government with adequate participation of states and union territories.

On the issue of interlinking of rivers a matter that has generated differing views and controversy about its desirability, an essay has been attempted by Raj Kumar Siwach. He refers to the Interlinking of Rivers (ILR) project, contemplated by the Government of India, to contain twin problems of drought and floods by way of transferring surplus water to deficient regions of the country through 30 major river links over 12,500 kms of canal length and construction of 32 dams. Although the project has yet to be finalized and implemented, its significance to the country's future water availability, usage and management cannot be underestimated.

The challenges that are likely to be confronted by this project pertain to civil society-state conflicts, inter-state differences in approach, ecological repercussions resettlement bottlenecks, legal wrangles, complex water governance problems and newer challenges for water administrators etc. Yet, the need for synthesizing equity, growth and innovative ideas, traditional knowledge and modern technology is paramount in this respect and according to Siwach, there does not seem to any alternative to this ambitious mega project. Interestingly, Siwach recommends restructuring the Central Water Commission into a multidisciplinary agency.

Drinking Water Availability and Supply in Urban Areas

Drinking water availability for the growing population has remained a critical problematic area of water management. Rakshat Hooja and Rakesh Hooja in their essay "Managing Water Supply in the 21st Century Metropolis: Lessons from Delhi" adopt a futuristic outlook while examining certain crucial issues pertaining to urban water supply.

It is estimated that by the year 2025, two-thirds of the world's population will comprise urban dwellers. In India, by 2020, about 50 per cent of population will be living in cities. No wonder, water will be a key resource for sustainable urban development. Provision of sufficient water for a multitude of necessities and prevention of pollution will emerge as a formidable challenge in the coming future. Already, urban water supply and sanitation in the country are suffering from inadequate levels of service, a challenging demand-supply gap, poor sanitary conditions and deteriorating financial and technical performance. In Delhi, India's capital, there is a niggling problem of constant water crisis, more so during the dry seasons. As the demand-supply gap widens, naturally, more groundwater is being exploited which is inimical to sustainable happiness of citizens.

Besides the regular government organizations, private players have entered the scene of water generation and distribution. In certain colonies of Delhi, Resident Welfare Associations (RWAs) have been able to influence water supply and its management. Hopefully, the role of these RWAs will expand and intensify in times to come.

The authors have identified seven major issues pertaining to water management in a metropolitan town like Delhi. These are: sourcing enough water; quality of water; methods of distribution; pricing of water; nature of water as a commodity/public good; role of private players in water supply; and role of citizen/consumer groups/resident welfare associations in water supply. All these issues need to be addressed while designing a new paradigm for water supply in a 21st century metropolis.

Hooja and Hooja opine that while planning for a mega city, we should keep in view the requirements of the main city habitation or municipal areas and its suburbs as well as attached peripheral semi-urban and rural areas. A growing city needs a futuristic plan that would take into account its growing size. Accordingly, water bodies being developed or restored in the proximate areas of the city, large water treatment plants and attached industrial and agricultural-horticultural use areas will form part of an expanded water management network. Rainwater harvesting will be a necessary component of water supply system, while large as well as a small treatment plants at intermediate points in the supply system will ensure good quality of water.

The water utility would cater to the larger urban agglomeration and develop expertise in urban water management, while the multipurpose municipal body covering a smaller area could be freed from the problem of water supply.

While the specialized public water utility system would manage the main storage treatment and supply system, water should be provided in bulk to community-based organizations or resident welfare associations.

The authors suggest an "out of the box" thinking in devising strategies for creation, distribution, conservation and recycling, and usage of water. These strategies have to be based on the maxims of rationality, prudence and foresightedness.

K.K. Gadeock's "Challenges and Priorities for Water and Sanitation Sector Reforms for Urban Areas" complements the preceding essay by Hooja and Hooja.

After briefly reviewing the present situation in India as regards water sources and sanitation vis-à-vis existing norms as also looking at cost recovery and at existing Indian models of delivery to small towns, and to metros and large cities, Gadeock endorses and spells out recommendations made by the World Bank, viz. (i) enhancing accountability of institutions to improve reliability of service, (ii) augmenting the financial sustainability of the service provider, (iii) promoting the environmental sustainability of service, (iv) improving the affordability of service, and (v) building the capacity of the WSS sector.

The latter part of the essay looks at the situation in Rajasthan and the role of the Public Health Engineering Department (PHED) as also at the available options for water supply in Jaipur city. Gadeock recommends following the examples of countries which have followed corporatization model with an effective regulatory agency, incentive-based contracts and private sector participation. As regards the government, he advocates that apart from institutional and organizational reforms, it will have to engage itself in critical areas such as groundwater management, establishment and management of water entitlements and rights, and the financing of public goods such as flood control and waste water treatment.

The Water Vision

Rakesh Hooja in his reflective essay on "Water Vision and Strategies at Global and All-India Levels" refers to the salient planks of the Water Vision prepared by the World Water Council. Everyone accepts that there is a water crisis today, but the crisis is not about scarcity only. It is also about the management of our limited water resources. Since water resource management and development is central to sustainable growth and poverty reduction, most developing countries need to be active in management as well as development of water resource infrastructure.

The Water Vision aims at empowering communities and other stakeholders and beneficiaries, producing more food and using water more productively, and managing water to conserve the quantity and quality of fresh water as also to ensure the existence of healthy catchments, encouraging greater innovation in water resource management and better governance. Accordingly, the key action areas envisaged by the Water Vision to meet these objectives include: involving all stakeholders in the process of facilitating integrated water management, ensuring full cost pricing of all water services, augmenting public funding for research and innovation, increased cross-national and inter-state cooperation in river basis and effecting massive increases in investment in water.

A few water policy reforms to enhance water security endorsed by Hooja are: establishing strong water rights and improving groundwater governance, improving incentives for efficient use of agricultural water, reforming policies to improve water quality, breeding and managing crops to conserve water and promoting research on the impact of global changes on water resources.

On the lines of the World Water Vision, the Indian Vision also aims at harnessing the untapped water wealth of the country, adding to the existing irrigation potential, ensuring fuller utilization of the created potential, effecting additional water storage facilities, improving the drainage system with a view to enhance productivity, optimizing water use as per agro-climatic conditions and bringing benefits of water resources development to water scarce areas through inter-basin transfer. To meet these objectives, the strategies suggested by Hooja include integrated management of water resource development, augmentation, proper utilization and preservation of the existing resources, encouraging participatory approach to water resource management, dissemination of technology and research, constructing water harvesting structures, adopting water harvesting practices and ensuring adequate allocation of financial outlays for the water sector.

Hooja's perspective is based on the premise that irrigation schemes cannot be considered separately from the area development schemes, but in fact they should to be treated as multidisciplinary, integrated and interdependent area development schemes that are based on irrigation but which also take into account occupational changes, fresh settlement or resettlement of populations and integrated development of sectoral programmes.

Above, then, are a few salient ideas incorporated in the present anthology covering multiple aspects, facets and dimensions of water management.

Chapter 2

Long-Term Water Demand Projections for India and Possible Policy Responses: A Note

Yoginder K. Alagh

THE Planning Commission of India has correctly projected that the net area sown or arable land of the country will remain constant at 141 million hectares. Growth in net area sown at around 1 per cent annual in the early period of planning fell to around 0.6 per cent and then to 0.3 per cent in subsequent decades and is now not growing at all. It is reasonable to assume that the geographical area of the country or the extensive land frontier for exploitation has reached its limits This is an important issue, the implications of which are not being realized with the urgency they deserve, since at a basic level resource constraints of a more severe kind faced by certain East Asian economies are now being approached in India. Organizations, communities, households and individuals will have to grasp this fact and live with it.

The intensive frontier for land use, however, remains. It has been known for example, as noted by Chaddha *et al.*, that cropping intensity depends on irrigation. Thus gross cropped area or harvested area has been shown in the past to be strongly determined statistically, in an econometric sense, by net irrigated area and irrigation intensity. Irrigation permits the possibility of multiple cropping by bringing additional land under cultivation and the same land to be used more than once. Also, the application of new technologies in the past was related to assured water supply. The new technology, on account of its photo insensitivity properties, permits shorter duration crops, which also is associated with increase in cropping intensity. (For details of this relationship in agricultural planning and policy models, see Alagh, ESCAP, 1983.) The use of this relationship has been used in Indian agricultural policy and plan models, since the mid-seventies when the first agricultural sub-model of Indian planning was formulated for grain self-reliance (see Alagh *et al.*, Planning Commission, 1979). The parameters used in different plans were as follows:

Table 2.1

Parameters Used in Different Plans

Plan	*Additional Irrigation Utilization (mn. hec.)*	*Additional Cropped Area (mn. hec.)*	*Elasticity of GCA w.r.t. GIA*
Fifth	9.11	6.04	0.20
Sixth	13.80	11.74	0.26
Seventh	10.90	10.00	0.31
Eighth	9.50	7.60	0.24

In the nineties as we noted arable area has stopped growing and so the land constraint is far more severe. Growth will now have to be sourced from double cropping and yields.

This fundamental relationship can be used to project the intensive resource base of the economy. Table 2.2 shows that by the end of the decade India would have used up most of its balance water reserves, with the irrigated area reaching around 114 million hectares by 2010 (see Alagh, 1995, p. 395 and table). The projections for 2020 are a requirement of irrigation of 122 million hectares for irrigation (Chopra and Golder, 2000, Table 2.6).

Table 2.2

Land and Water Resources in Perspective

Variable	*1991/2*	*1996/7*	*2001/2*	*2006/7*
1. Population (millions)				
a. Planning Commission**	856	938	1016*	1099
b. UN (unrevised)	874*	955	1042	1130*
2. Net Area Sown (mn. hec.)				
a. Planning Commission estimate	140	141	141	141
b. Revised		141	141	141
3. Gross area sown (mn. hec.) estimate				
a. Planning Commission	182	191	197	203
b. Revised	183	191	197	205
4. Gross Irrigated Area (mn. hec.) estimate				
a. Planning Commission	76	89	102	114
b. Revised	6	8	92	107
5. Cropping Intensity				
a. Planning Commission estimate	1.3 0	1.3 5	1.4 0	1.4 4
b. Revised	1.3 0	1.3 5	1.4 0	1.4 5
6. Gross Irrigated Area as % of Gross Area Sown				
a. Planning Commission estimate	41.5	46.9	51.7	56.1
b. Revised	35.0	41	46	51

Source: Perspective Planning Division, Planning Commission FAO, Agriculture Towards 2010, Rome.

Revised projections are tentative and are by the author.

Note: * Interpolated or extrapolated from implied trends.

** Planning Gommission estimates.

Source: Uma Lele, *et al.*, World Bank, 2001, Annex table by Y.K.Alagh.

The projections assume a vastly improved performance on the land and water management frontiers. It needs to be remembered that the balance ground water reserves are now more limited. A very dramatic effort will be needed to harvest and carefully use the available water. Otherwise, the projected increase in cropping intensity will simply not take place. Cropping intensity increased from around 1.18 at the beginning of the seventies to around 1.3 in the early nineties. In the next two decades, this effort needs to be considerably strengthened, so that cropping intensity can increase from 1.3 to 1.5. Harvesting of rainwater, recycling water from agricultural drainage systems, more judicious use of water for cropping, will all be required. Non-agricultural use of water will have to be far more economical. The detailed exercise done for this study requires that in the sustainable scenario 35.83 BCM of water are saved by conjunctive use of surface and groundwater and 142 BCM through harvesting of runoff (Chopra and Golder, *ibid.*, Table 2.6).

Another way of looking at the severe land constraint is to see that a net area sown per person will go down from around 0.17 hectare to around 0.10 hectare. Gross area sown per person currently around 0.2 hectare will, even if cropping intensity increases very rapidly, go down to around 0.15-0.18 hectare.

Table 2.3 given below summarises the findings on water requirements and water availability under alternative scenarios. Extrapolations of demand from different sectors show that if business as usual continues, quantitative shortages of water are likely to emerge. Declining water use efficiency in agriculture, increasing urbanization and unregulated industrialization pose significant challenges for the water sector in the future. Shortages, either of ground or surface water or both, are likely to be pronounced in the states of Andhra Pradesh, Gujarat, Haryana, Punjab, Tamil Nadu and Maharashtra.

Table 2.3

Water Requirements: Different Scenarios

(in BCMs in 2020)

	BAU	*HG*	*SS (%)*
Households	67.52	67.52	45.01 (4.66)
Power	8.19	12.29	5.00 (0.5)
Industry	27.91	41.58	27.72 (2.87)
Agriculture	677.30	804.20	768.37 (79.69)
Evaporation	42.00	42.00	42.00 (4.33)
Ecological	78.00	78.00	78.00 (8.09)
Total	920.92	1005.59	964.09

Even with this a shortfall of irrigated land to the magnitude of 10 million hectares may arise. If this shortfall is made up for as in the HG scenario, a water shortage or deficit of 22 per cent arises. The manner in which this translates into groundwater or surface water shortages in particular regions depends on policies pursued. The HG scenario also increases demands of the industry and power sectors, resulting in an overall increase in water requirement.

The sustainable scenario identifies interventions on the demand management and supply augmentation sides that can ensure that total water requirement is 964.09 BCMs in 2020. Of this, 79 per cent shall come from the agricultural sector. In percentage terms, this is a decrease from current levels since requirements for non-agricultural sectors rise with industrialization and urbanization. Only 4.66 per cent of total requirement comes from the household sector and another 3.37 per cent from industry and power sector. This study provides additionally for a requirement of 78 BCM to maintain base non-seasonal flow in rivers and 42 BCMs for evaporation losses. With total supply from ground and surface water at 1110.566 BCMs, one can argue that the position at the aggregate level shall be manageable. Such a presumption assumes however that interventions suggested to achieve improved water use efficiency shall be undertaken and shall be successful. In frozen water use efficiency in agriculture scenario, acute shortages may arise even at the aggregate level.

By 2020, Chopra and Goldar (*ibid.*) estimate a BAU scenario as under:

> In such a scenario, overall water shortage or deficit is only of 2 per cent. This is accompanied by an under-utilization of surface water capacity of 21 per cent (due to low water use efficiency) and an over-extraction of ground water of 25 per cent. Such an unbalanced growth shall itself be the source of a considerable amount of unsustainability.

The second interpretation of the BAU scenario is motivated by the need to estimate regional shortages or surpluses. The BAUST or the business as usual with the state level estimates is made by assuming that surface and ground water development follow the trend extrapolated from the past.[1] The regional analysis reveals that over-extraction of groundwater shall emerge in eight states. In addition to Gujarat, Haryana, Punjab and parts of Uttar Pradesh which are characterized as areas with overuse of groundwater, Andhra Pradesh, Maharashtra and Tamil Nadu are also expected to be subject to over-extraction. Surface water shortages may start emerging in some states such as Gujarat, Bihar, Maharashtra and Orissa, and again taking population projections and the energy models as developed by the AITD urbanization study and suitable assumptions they estimate:

> The total requirement for households, power and industry is 103.62 BCMs in the BAU scenario. Adding evaporation loss and ecological requirement we obtain a total requirement of 223.62 BCMs for requirements other than agriculture.

Chopra and Goldar (*ibid.*) also estimate a high growth scenario. Then:

> The high growth scenario at the national level implies a rate of growth of agriculture of 4.98 per cent per annum with the food grain sector growing at 2 to 2.4 per cent p.a. and the non-food grain sector at 7.69 to 9.22 per cent p.a. Irrigated land requirement increases to 122 million hectares. With present levels of water use efficiency, water requirement increases to 804.2 BCMs. This implies a shortfall/water deficit of 22 per cent. The manner in which this translates into over-extraction of groundwater or other indices of unsustainability depends on policies pursued and cannot be ascertained. It is, however, clear that both the high growth and BAU scenarios are likely to result in unsustainable demand for water of one kind or the other in the absence of specific investments and policies directed at improvement of water-use efficiency and other sustainability promoting measures.

The requirements of the households, power and industry sectors likewise add up to 120.57 BCMs in the HG scenario as against 103.62 BCMs in the BAU scenario. Regarding the quality of water, they say:

> An important issue which our analysis brings out is that at present levels of urban wastewater and effluent treatment, water quality will deteriorate significantly, impacting availability as well in certain areas. The reduction in water availability due to quality problems is expected to be substantial.[2] We use this evidence to

underscore the need for and the corresponding benefit from 'sustainability investments'.

Chopra and Goldar (*ibid.*) then develop a sustainable policy scenario:

> It is clear from the above analysis that, in order to achieve sustainable water development, consistent with sectoral requirements arising from a sustained 7 to 8 per cent annual growth in GDP and other accompanying changes in the economy, a large number of interventions shall be required. These interventions could be technological, institutional, or supply augmenting possibilities. Among some interventions for sustainability we suggest, including watershed management, drip irrigation, and institutional arrangements, such as water users' associations, that help in more efficient use of water. Some of these have been experimented with in different contexts and we estimate (on the basis of existing studies) the extent to which they can be expected to spread over the next twenty years. Corresponding costs are also worked out. These are, in other words, the costs to be borne for ensuring sustainable use of water. Other interventions suggested involve costs which are either difficult to measure, involve uncertainty or are not only economic costs but also perhaps political in nature. Further, interventions are also classified according to whether they are to be undertaken on presently irrigated land which is privately owned, or consist of interventions on land (possibly under private, common or government ownership) to be brought under irrigation.

Table 2.4

Nature of Interventions under Sustainable Scenario

Nature of Interventions	*Sector*	*Water Saving/Addition (in BCMs)*
Demand Management/ Recycling Technology	Households	22.51
	Power Sector	7.29
	Industry	13.86
	Agriculture	35.83
Supply Supplementing	Additional Runoff Capture	142.00
Total Water Saving/ Augmentation		221.49

Much the same kind of approaches and projections have been made by the National Commission on Perspectives for Water Development as shown below in Table 2.5:

Table 2.5

Water Requirement for Different Uses

S. No.	*Uses/Year*	*2010*			*2025*			*2050*		
		Low	*High*		*Low*	*High*		*Low*	*High*	
		km³	*km³*	%	*km³*	*km³*	%	*km³*	*km³*	%
	Surface Water									
1.	Irrigation	382	391	53	360	389	46	375	463	39
2.	Domestic	23	24	3	30	36	4	48	65	6
3.	Industries	26	26	4	47	47	6	57	57	5
4.	Power	14	15	2	25	26	3	50	56	5
5.	Inland Navigation (in additional for ecological need)	7	7	1	10	10	1	15	15	1
6.	Environment Ecology	5	5	1	10	10	1	20	20	2
7.	Evaporation	42	42	6	50	50	6	76	76	6
	Total	499	510.1	70	532	588.3	67	641.1	751.7	64
	Ground Water									
1.	Irrigation	184	188	26	211	299	27	253	344	29
2.	Domestic & Municipal	19	19	3	25	28	3	42	46	4
3.	Industries	11	11	2	20	20	2	24	24	2
4.	Lower	4	4	1	6	7	1	19	14	1
	Total	217.7	221.9	30	262.3	281.7	33	331.9	428.3	36
	Grand Total	717	732	100	794	850	100	973	1180	100

Source: Government of India, National Commission on Perspectives for Water Development.

Assuming very optimistic policies, the Commission on Water Perspectives, which followed the approaches indicated above, has projected balanced demands/supplies in 2025. These include:

- Improvement in irrigation efficiencies above 60 per cent from existing levels of less than 40 per cent.
- Conjunctive use and aquifer management.
- Demand management and rationalization of cropping patterns.
- Much greater priority to resource allocation to the water sector.
- Improved project management and phasing.
- Watershed development.

But, separately, its scholarly chairman has projected a shortage of 25 per cent in a trend forecast. A recent UNU study by Y.K. Alagh and Kirit Parikh (2000) has also projected problems and at the regional level worse.

While some of the interventions suggested are well known, others involve fresh thinking, hence we discuss them in some detail, with theoretical and experience based analysis.

The technology interface is important, both for land and water management and for cropping and non-crop farm systems that are optimal, in this class of issues. While a lot of research has been done and is available (Alagh, FAO/UNESCO, 2002) the real issues are policy rules for fast replicability of existing knowledge and success stories. Community institutions have to be at the heart of this process. The projects examined have varied considerably. Watershed development, for settled agriculture alternately tree crops, reclamation of saline lands, farmers run lower level irrigation systems, aquifer management in difficult situations, like coastal aquifiers, tribal irrigation cooperatives, tank irrigation have all been reported as success stories and studied. The question is replicability on a larger scale. We have (Alagh, 2003) tried to set out some policy rules which, we argued, if applied in functioning policies may reverse the tide. These problems were the basis of the ACRP project which gave requirements at an agro-climatic zonal level.

These plans include those built up by community groups, NGOs and cooperatives. The Agro-Climatic Project has identified the major components:

- Farmer-led distribution systems in existing commands; watershed development rehabilitation of tanks, beels, ponds, talaavs; aquifer management particularly in coastal areas have high returns.

- A massive programme for land and water development has to be the centre piece. Growth of private investment in groundwater has been spectacular and needs further support, particularly in the eastern region of India.
- Basin development and transfer of water are important.
- It must be led by a public policy initiative, with a very large financing component. Newer institutional forms like cooperatives, NGOs, non-profit companies and others must be designed.
- Leadership groups must supervise the preparation and implementation of such programmes at the regional level of three to five districts.
- Conjunctive water use can be attempted in an area and is well suited to describe the "dynamic interaction between the surface and subsurface water systems" (Millington, 1996). The flow chart of a simple multi-level model can be as follows:

```
•   Water ------→ Population; Needs, irrigation, trees, etc.
•   Development                                          \
•                         \                              \
•                         \                              \
•                         \                              \
•                         \                              \
•                         \                        Wells Drilled
•                         \                              \
•   Climate Regime ---→ Water Table                      \
•   Depth --------------→ Deep Aquifer Water Quantity
•                         \                              /
•                         \                              /
•                         \                              /
•            Ground Water Quantity
```

Source: Nicholas Sonntag (1996).

It is interesting that in recent global meetings the same strategy is being advocated. India is playing a strong role in such advocacy. For example, the Expert Round Table organized by

India and the UN to operationalize the new initiatives required at the Johannesberg meeting on RIO PLUS 10 said the following:

- *Improve* investment processes in developing countries and countries with economies in transition to facilitate access to credit lines as well as to preferential terms of financing and of providing funds for collateral support systems and sharing of investment risk. In this context, provide securities for local institutions involved in infrastructure development and specific knowledge based activities to support sustainable economic growth, through, for example, creation of collaterals, interest differentials and trading of financial papers. These processes should be targeted, amongst others, to artisan and producer groups linked with local and global markets, local government agencies providing social and economic infrastructure, and farming and rural communities.
- *Improve* coordination among international financial institutions and redirect funds to sustainable development projects.
- *Develop new or strengthen* existing mechanisms such as the Clean Development Mechanism (CDM), to finance or re-finance community projects in rural areas aimed at land and water development, agricultural diversification and agro-processing, development of infrastructure, trade, and rural energy supply.
- *Study* for the purpose of replication, existing models for providing access of rural communities to ICTs in order to enhance the level of information in rural communities on productions, crops, markets, prices and technologies.

It has been shown recently that such work is important in sustainable development policies for the subcontinent (Alagh, 2004).

■ Notes

1. Following such a methodology implies that the two estimates should not be compared. The BAUST estimate is motivated by the need to identify regional shortages or surpluses of water, while the BAU estimates determine the impact of food security objectives on water demand.

2. See Das and Dipankar (2000) using a more stringent standard for irrigation water. This study estimates that in some areas with a high level of urbanization, only 54 per cent of the groundwater in underlying aquifers shall be fit for use for irrigation. This study was commissioned by IEG as part of the current UNU study.

■ References

. Alagh, Y. (Chairman) 1979, *Task Force on Minimum Need and Effective Demand*, Government of India.

Alagh, Y. *et al.*, 1979, *Studies on the Structure of the Indian Economy and Planning for Development*, PPD, Planning Commission, New Delhi.

—, *et. al*, 1984, *Policy Modelling for Planning in India*, in E. Thorbecke, ed., below.

—, 1988. Guidelines for Agroclimatic Planning: A Draft for Discussion, *Journal of Land Development.*

—, 1991, *Indian Development Planning and Policy.* WIDER Studies in Development Economics, Helsinki and Delhi, Vikas.

—, 1991, *Sustainable Development, From Concept to Action: Techniques for Planners,* UNCED.

—, 1994, *Planning and Policies for Indian Agricultural Research*, 25th Lal Bahadur Shastri Shastri Lecture, reprinted in ICAR, Landmarks in Indian Agriculture.

—, *Next Phase of Agroclimatic Research,* in ARPU Vol. 2 of same title, New Delhi, Concept.

—, and D. Buch. 1995, *The SSP and Sustainable Development,* in W. Fisher, Columbia University Seminars, below.

—, 2000, *Sustainable Development: India 2020,* Tokyo. UNU/IAS.

—, 2003, Emerging Institutions for Rural Development, in Chopra, K. C.H.H. Rao and R.P. Sengupta, eds., *Water, Sustainable Livelihood and Eco-System Services,* Concept, Delhi.

—, 2004, Our Common Future, Working Paper, Kieo University 21st Century Programme, Japan.

Bhalla, G. and Y. Alagh, *Performance in Indian Agriculture,* New Delhi, Sterling.

Chopra. K. and G. Kadekodi, 1993, Watershed Development, *Economic and Political Weekly,* June 26, pp. A61-A67

Chopra, K. and Goldar, B., 2000, *Sustainable Framework for Water,* UNU, IAS.

Frankel, FR., 1971, *India's Green Revolution: Economic Gains and Political Costs,* New Jersey: Princeton University Press. 1971.

Griffin, Keith, 1971, *Green Revolution: An Economic Analysis,* United Nations Institute for Social Development, Report No. 72.6, Geneva.

Lele, U.N. Kumar, Y. Alagh, N. Saxena and K. Mitra, 2000, *Forestry in India: An Evaluation,* Washington, World Bank.

Nasution, L., 1993. *Agricultural Regionalisation of Indonesia,* Bogor, Agricultural, Research Centre.

Sachs. 1., 1991, *Background Paper for the Hague Declaration,* extensively reprinted in Nature and by RIS.

Sadasyuk, G. and P. Sengupta, 1968, *Economic Regionalisation of India.. Problems and Approaches,* New Delhi, Census of India, Monograph No. 8 of Census, 1960.

Sonntag, N., 1996, *Adaptive Management Policy Exercises: Two Methods for Resource Management,* in MRC, p. 27.

Chapter 3

Water Savings and Conservation in Irrigated Agriculture

S.A. Kulkarni

RAPIDLY growing population, increased economic activity coupled with improved living standards has led to enhanced competition for and conflicts over the limited fresh water resources. During the 20th century, the world population has increased three-fold whereas water withdrawals have increased by about six-fold. It is estimated that currently one-third of the world's population lives in countries that experience medium to high water stress. This ratio is expected to grow to two-third by 2025. Growing water scarcity has led to revised thinking in the way the water resources are developed and managed. Globally, about 1500 million hectares are under arable and permanent crops (APC), of which some 270 million hectares (18 per cent of APC) are irrigated. Irrigated agriculture contributes 40 per cent of all the food production in the world (FAO, 2001). The International Commission on

Irrigation and Drainage (ICID, 2000) estimated that current food production would have to be doubled within the next 20 years for which there is a need for global increase of withdrawals of 15-20 per cent of waters from surface and ground for providing irrigation to larger areas. The Food and Agriculture Organization (FAO, 2002) anticipates a net expansion of irrigated land of about 45 million hectares in 93 developing countries and projects that agricultural water withdrawals will increase by some 14 per cent from 2000 to 2030 to meet future food production needs.

An IWMI-IFPRI research study (IWMI, 2004) on global water outlook to 2025 has revealed that the challenge posed for food production can be addressed through three broad strategies: (1) investing in infrastructure, (2) conserving water and improving the efficiency of water use, and (3) improving crop productivity per unit of water and land. If these are incorporated into water saving strategy, the research argues that it will be possible to avert the growing water crisis.

The terms 'water savings' and 'water conservation' are generally associated with the management of water resource under scarcity. In irrigated agriculture, water savings aims at limiting or controlling water demand and use and minimizing of water wastes and misuse of water. Water conservation refers to all those policies, measures and practices that aims at conserving or preserving the water resource, as well as combating the degradation of water resource including its quality (Pereira *et al.*, 2002).

India's Land and Water Resources, Population and Food Production

India has 2.45 per cent of world's land, 4.9 per cent of fresh water resource and 17 per cent of population. From the reported area of 306 million hectares for land utilization statistics, the present cultivated area is 142.5 Mha (47%), forested 69 Mha (23%), and 42 Mha (14%) is not available for cultivation (Ministry of Agriculture and Co-operation, GoI, 2004). Out of the cultivated area, 57 Mha (40%) is irrigated which produces 55 per cent of food, 60 per cent is rainfed area and produces 45 per cent of food. Amongst the net irrigated area, about 17.7 Mha (31%) from canal water, 33 Mha (58%) from groundwater and balance 6 Mha (11%) from tanks and other sources (CWC, 2003). Ultimate proportion of irrigated and rainfed area is estimated to be 55:45 producing 75:25 of required food (NCIWRD), 1999). In future, more and more land will be occupied by urban habitats and industries. Some land will go under submergence of reservoirs and for rehabilitation and resettlement of displaced people from different developmental projects.

Water Resources Development

India is gifted by nature with the large number of major and medium rivers and the country is divided into 24 river basins. The water resources of the country are shown schematically in Figure 3.1 Total annual renewable water resource of the country is estimated at 1,953 km^3. The potential utilisable surface water resource is considered to be 690 km.^{3} The utilisable component assumes the provision of feasible storages but excludes major inter basin transfers. The recharge to ground water is estimated as 432 km^3 of which 396 km^3 is considered as potentially utilisable and includes estimated return flow of 90 km^3 from the present utilization. Thus, the total annual utilisable water is estimated as 1,086 km^3. Present utilization of water resources is about 629 km^3 (about

Figure 3.1

Water Resources of India

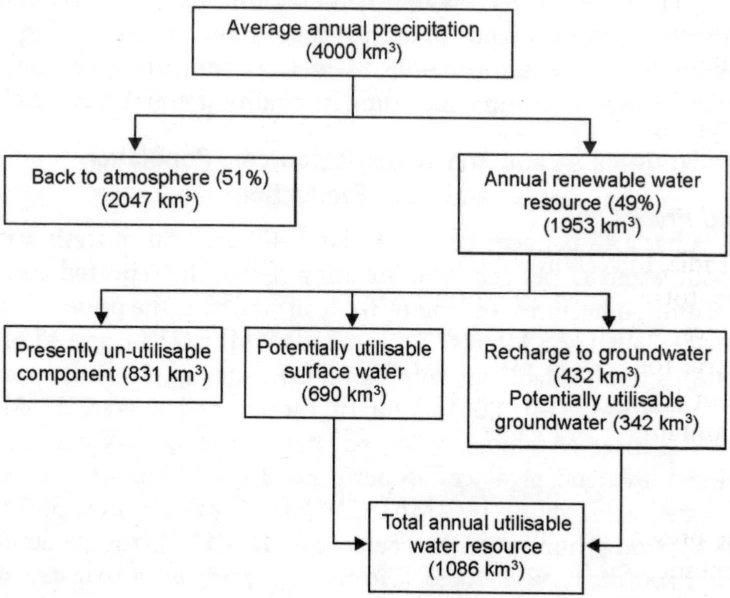

58% of the utilisable). It is estimated that of the potentially utilisable surface water, 384 km^3 can be used through creating storages. At the end of the year 2002, 4,525 dams were constructed creating a total live storage capacity of about 175 km^3. It is envisaged to create an additional 75 km^3 of storages by 2015.

Population Growth

India's population increased from about 360 million in 1951 to 1027 million in 2001. This resulted in decline in the per capita availability of agricultural land from about 0.4 hectare in 1951 to 0.14 hectare in 2001. Average water availability per capita per year has declined from about 6,000 m^3 to 1869 m^3 leading to the water scarcity mark of 1,700 m^3. However, in some basins like Sabarmati, Pennar, the per capita availability is even less than 1,000 m^3 indicating water stress situation. Although urbanization rate in India is among the lowest in the world, yet urban population is second largest in the world in absolute numbers. India therefore needs urgent measures for population control. China has brought down the population growth rate to 1 per cent by adopting the two-child norm in the 1990s and enacting the one-child population control law in 2002, while India's population continues to increase at 1.9 per cent per annum. China is close to stabilization of its population while India's population is likely to stabilize at 1,600 million by 2050. It is therefore necessary to focus on water resources development and management strategy to meet the water needs of food, people and eco-system at that level of population

Food Production

In India, foodgrain includes rice, wheat, gram and pulses together. The total foodgrain production of the country is 212 metric tonnes, which includes about 71 metric tonnes of wheat and 85 metric tonnes of rice. Gross per capita foodgrain availability is about 160 kg/year or about 450 gram per day, which is below the minimum recommended nutritional level of 575 gram.

Sector-wise Water Demand and Requirement

The present demand and future water requirement for irrigation, domestic, industry, power, navigation and ecology as assessed by the National Commission on Integrated Water Resource Development (1999) are depicted in Table 3.1. Presently, in India, about 83 per cent of water is withdrawn for irrigation, 4 per cent each for domestic and industrial use, and 2 per cent in power sector. It is estimated that by 2025 and 2050, the share of irrigation water in total water withdrawals will decline to 72 per cent and 68 per cent respectively. The total annual demand for all uses by year 2025 is likely to be 843 km^3 (43% of the renewable water resource) as against 629 km^3 in 1998. The country is likely to confront water crisis in the year 2050, as the total water requirement would be 1,180 km^3 (60% of the renewable water resource) that is higher than the total annual utilisable water resource (1,086 km^3) Generally, it is considered that if the total water withdrawals exceed 40% of the total annual renewable water resource, it indicates water stress situation, while between 20% and 40% indicates impending water scarcity. On this basis, India will have reached the water stress situation by 2025 (WWDR, 2003).

Table 3.1

Sector-wise Water Requirement in India

km^3

Sector/Year	*1998*	*2025*	*2050*
Irrigation	524 (83)	611 (72)	807 (68)
Domestic	30 (4)	62 (7)	111 (9)
Industry	30 (4)	67 (8)	81 (7)
Power	9 (2)	33 (4)	70 (6)
Navigation	0	10 (1)	15 (1)
Ecology	0	10 (1)	20 (2)
Evaporation/Losses	36 (7)	50 (6)	76 (7)
Total	629 (100)	843 (100)	1180 (100)

Note: Figures in brackets indicate percentages.

Challenge and Policy Options

Major challenge for the country is to provide needed water, both in quantity and quality to achieve food sufficiency and environmental sustainability keeping in view the projected population in the year 2050. It is argued that this challenge can be met by (1) integrated water resources development and management (IWRDM); (2) improving irrigation efficiency; and (3) improving water productivity.

Integrated Water Resources Management (IWRM)

Global Water Partnership (2000) has defined the IWRM as "the process, which promotes the co-ordinated development and management of water, land and related resources, in order to maximize resultant economic and social welfare in an equitable manner without compromising sustainability of vital eco-system." Integrated management means that all the different uses of water resources are considered together. It is different from sectoral approach. However, in practice, there is very little integrated water resources management on the ground. Integration could be between:

- Land and water management
- Quantity and quality in water resource management
- Rainfed and irrigated agriculture
- Large-scale and small-scale irrigation
- Surface and groundwater
- Public and private schemes
- Upstream and downstream users
- Different water user agencies/departments/organizations/ministries
- Food, people and environment sectors
- Blue water and green water
- Fresh water management and coastal zone management
- Fresh water management and waste water management

During the recent past, an important conceptual advance has been introduced in water resource, i.e., blue water and green water. Blue water is that which flows in rivers, stored in reservoirs and recharged ill aquifers, while green water is that component of rainfall which is stored in 8011 profile/aquifer and is used directly for biomass production and lost in evapo-transpiration before contributing to runoff. Most water management generally focuses on the blue water, thus neglecting rain and soil water management. Management of green water holds significant potential for water saving both in rainfed and irrigated agriculture. Rainfed agriculture uses large volume of green water compared to irrigated agriculture. Thus, it is more meaningful to look holistically at the hydrologic cycle and consider terrestrial rainfall as the primary water resource.

In a water assessment study of the Sabarmati basin (ICID, 2004), it was found that irrigation withdrawals were 93 per cent of the total withdrawals for all use sectors. However, of the total consumptive use for the irrigated agriculture, 50 per cent was satisfied by rainfall while balance 50 per cent was supplemented through irrigation.

Improving Irrigation Efficiency

Classical irrigation efficiency is usually defined as a ratio of water volume beneficially used by plants to volume of water delivered at a water source. However, the perception of "efficiency" differs amongst engineers, agronomists and economists. Engineers compare the volume of water delivered and consumed, agronomists measure the crop output (yield) per unit of water delivered, and economists compare the value of output and the opportunity cost of water (Wichelns, 1999).

Recently, it has been debated that the classical concept of irrigation efficiency is misleading. Conventionally, irrigation efficiency considers water flowing into drain or deep percolated below the root-zone of the crop as a "loss". But if these losses are used by farmers further downstream, either directly or through recharge of aquifers, then improving the on-farm efficiency may not lead to increased efficiency at river basin level. In other words, it implies that while efficiencies at field level in a river basin may be low, the over all efficiency of the basin may be quite high, and the "real" opportunities to save water by improving "irrigation efficiency" may be limited. This will be particularly applicable in case of "closed" basins, where all useable return flows are captured and reused. However, it should be remembered that these water losses might cause adverse environmental impacts like waterlogging and salinity and degradation of water quality besides increasing pumping cost.

At a basin level, what may look like a wasteful use from a crop irrigation perceptive, may be beneficial for another type of use. New concept to distinguish between consumptive and non-consumptive uses, beneficial and non-beneficial uses, and reusable and non-reusable fractions of the non-consumed water diverted into an irrigation system or sub-system is being proposed. This concept appears to be more relevant than – irrigation efficiency" for the formulation of water savings and conservation policies and measures.

It is therefore necessary to quantify the proportion of water use i.e. diverted for given use, that is consumed and the proportion that is not consumptively used and is available for reuse or becomes degraded after use. Figure 3.2 shows the consumptive and non-consumptive components of irrigation water diverted to irrigation system. In this context, improvement of irrigation efficiency is of great importance under water scarcity so as to

Figure 3.2

Consumptive and Non-consumptive Components of Irrigation Water at a Basin Level

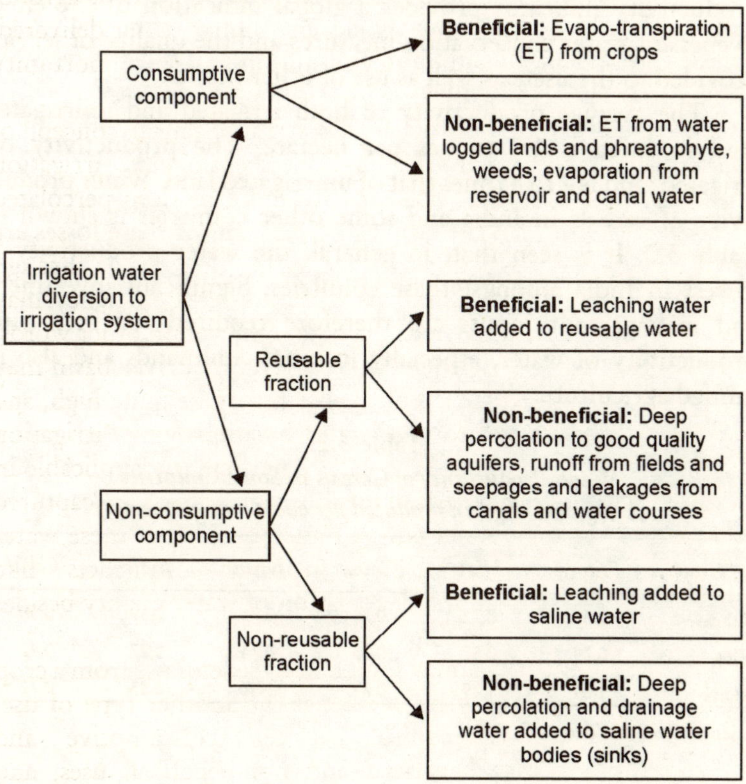

increase the beneficial use of water for crop production. Thus, water saving can be achieved by reducing non-beneficial ET and lessening non-beneficial reusable fraction and non-reusable fraction of non-consumptive component of irrigation water diversion.

Improving Water Productivity

According to Molden *et al.*, (2004), by improving productivity of water used for agriculture, it is possible to reduce the amount of additional fresh water withdrawals needed to feed the world's growing population to zero. This appears to be a too ambitious goal. The term "water productivity" (WP) expresses the quantity of product or value of the product per unit of water. In an irrigation scheme, it may happen that even though the irrigation efficiency is very high, water productivity could be low. Water productivity indicator provides a global indication of the effectiveness of water conservation measures and the quality of service provided to the users as well as use of other inputs.

The average productivity of both irrigated and unirrigated land in India is 1.7 tonnes per hectare. The productivity on irrigated land is 2 to 4 times that of unirrigated land. Water productivity of cereals in India and some other countries is shown in Table 3.2. It is seen that, in general, the water productivity is lowest in India amongst these countries. Significant investment and management inputs are therefore required for increasing productivity of water, especially in canal commands and also in rainfed agriculture.

Table 3.2

Water Productivity of Cereals in Some Countries (kilograms of grain produced per cubic metre of water)

Crop	*France*	*USA*	*China*	*Japan*	*India*	*World*
Wheat	1.52	0.72	0.78	0.74	0.39	0.56
Rice	0.79	0.52	0.73	0.74	0.27	0.42
Maize	1.64	1.49	0.84	NA	0.23	0.72

Adapted from Molden *et al.*, (2004).

Water Savings and Conservation Initiatives and Measures

International organizations like International Commission on Irrigation and Drainage (ICID), Food and Agriculture Organization (FAO) and International Water Management Institute (IWMI) have already taken up initiatives to support and promote water savings and conservation in agriculture. ICID launched its global water savings (WatSave) programme by setting up the WatSave Work Team in the year 1999 with the objective of promoting and recognizing water savings/conservation practices amongst its member countries. ICID has been compiling and disseminating information on water savings/conservation practices adopted world wide through publications, website, organizing regional workshops and through annual WatSave Awards (www.icid.org). An internet-based information system on water conservation and use in agriculture called WCA-InfoNET is provided by IPTRID/FAO. The site <http://www.fao.org/iptrid/infonet/index.htm> allows direct access to publications, documents, data, computer programmes and discussion groups which provide a knowledge base and supports the necessary global platform for decisions on water conservation and use in agriculture. IWMI had launched the Comprehensive Assessment of Water Management in Agriculture by bringing together scientists, policy makers, development professionals and water users worldwide to find out potential tools for increasing water productivity in rainfed and irrigated agriculture.

As is evident, rich knowledge base, well-known practices and tools for water savings and conservation are available but their application is far from satisfactory. Summary of often-referred engineering, agronomic and management measures is given in Box 3.1. The adoption of water conservation and savings is not only a technological issue but involves many other considerations like social, economic, legal, institutional, policy and management decisions.

Concluding Remarks

India needs to adopt the strategy of integrated water resources development and management (IWRDM) to cope up with the increasing water demand from irrigation, domestic and industrial

Box 3.1

Water Saving and Conservation Measures in Irrigated Agriculture

Engineering

- Improving irrigation efficiency (conveyance, distribution and onfarm application)
- Re-modelling/modernization of the canal and control structures
- Construction of en-route water storages
- Use of precision on-farm irrigation methods – laser land levelling, micro and sprinkler irrigation
- Use of pipes for conveyance and distribution of water

Agronomic

- Improving water productivity on both rainfed and irrigated lands
- Use of poor quality waters (waste water, saline water, drained water)
- Dry seeding, raised beds, alternate wetting and drying for paddy crop
- Adopting deficit irrigation
- Use of mulches/control of weeds
- Crop breeding, use of early maturing crop varieties
- Switching from high-water consuming crops to less-water consuming crops
- Adopting drought and salt resistant crop varieties
- Practicing on-farm irrigation scheduling
- Improving non-water inputs (e.g., organic fertilizer, pesticides)
- Upgrading of rainfed agriculture by supplemental irrigation

Managerial

- Integrated water resource management (IWRM)
- Improved management, maintenance and operation of irrigation systems
- Benchmarking of irrigation schemes and water auditing

Contd...

Contd...

- Sustainable pricing of canal water and energy supply for pumping of groundwater
- Participatory irrigation management
- Developing and implementing water laws, institutional and policy reforms – river basin organizations
- Involvement of women in irrigation
- Training/capacity building from farmers to government officers/staff
- Promoting mass scale public awareness and involvement of stakeholders
- Strengthen/revamp irrigation extension services
- Use of media in propagating water savings/conservation
- Updating of water resource database and use of decision support and expert systems
- Promoting vegetarian diet

sectors keeping in view the population growth by 2050. The present irrigation development has lead to some negative impacts like waterlogging, salinity and over-exploitation of groundwater, adversely affecting the eco-system. It is essential to understand clearly consumptive, non-consumptive, beneficial and non-beneficial components of water use at project/sub-basin/basin level. Consequently, identify places where real water savings and conservation opportunities exist and the measures to achieve those. The project authorities should develop a comprehensive water saving/conservation strategies at various scales – from field to basin level and implement it aggressively to achieve tangible results. Lessons can be learned from worldwide experiences. There is a definite need to significantly improve the present irrigation efficiency and water productivity. Existing water management institutions need incentives to be more efficient and offer services that will encourage improvement in water productivity. Farmers need incentives to adopt technologies for water savings.

■ References

Central Water Commission (2003): *Pocket Book on Water Data*, CWC, MoWR, India.

Food and Agriculture Organization (2002): *World Agriculture: Towards 2015/2030.*

FAO (2001): *FAO Yearbook: Production*, Vol. 55, FAO Statistics Series No. 170, FAO, Rome 2002.

FAO, IPTRID website (http://www.fao.org/iptrid/infonet/index. htm).

Global Water Partnership (2000): *Integrated Water Resources Management*, Technical Advisory Committee (TAC). Background Papers, No. 4.

International Commission on Irrigation and Drainage (2004): Report on Water Policy Issues: India Country Study – Basin Level Assessment and National Consultation; Country Policy Support Programme, (www.icid.org).

International Commission on Irrigation and Drainage (2000): Strategy for Implementation of ICID's Concerns Emanating from the Sector Vision on Water for Food and Rural Development (Draft Report) (www.icid.org).

International Water Management Institute (2004): *Annual Report 2002-2003.*

Ministry of Agriculture and Co-operation, Government of India (www.agricoop.nic.in).

Ministry of Water Resources, Government of India (1999): *Report* of the National Commission for Integrated Water Resources Development (Vol. 1).

Molden David and Fraiture Charlotte de (2004): *Investing in Water for Food, Ecosystems and Livelihoods*, Blue Paper, Stockholm Water Week, Comprehensive Assessment of Water Management in Agriculture, IWMI, Sri Lanka.

Pereira Luis S., Cordery Lan, and Lacovides Lacovos (2002): *Coping with Water Scarcity*, International Hydrological Programme, IHP-VI, Technical Documents in Hydrology, No. 58, UNESCO, Paris.

Wichelns Dennis (1999): Economic Efficiency and Irrigation Water Policy with an example from Egypt, *International Journal of Water Resources Development*, Vol. 15, No. 4, pp. 543-560

World Water Development Report (2003): *Securing Food for a Growing World Population*, World Water Assessment Programme (WWAP), UNESCO (www.unesco.org/water/wwap/targets).

Chapter 4

Managing Water in Rural Areas

An Institutional and Multi-Use Perspective

Rakesh Hooja

THE pressure on water resources is growing day by day. Populations are increasing. Lifestyle changes have led to the per capita demand for water rising. As societies move from the earlier agrarian to the now more complex, multi-sector, societies and as technologies open up new possibilities of economic activity, the requirements of water grow. Many of the water sources – surface or groundwater – which were comparatively easier to tap have already been developed and are being exploited, and the development of new water sources now proves more costly and may involve complex technical and environmental problems which need to be sorted out before the sources can be tapped. Gradually, agitations and local competition, and even conflicts, over issues of water sharing have started emerging in our country.

Whether it is storage and distribution of water through canals, or pumping of ground water, or rainwater harvesting and watershed development, all have their differing challenges to be overcome.

Thus, the mix of water management and water development strategies and practices that are being adopted have assumed great contemporary importance. In fact, the entire issue of governance in the water sector, particularly at district, project and local levels, needs to be looked into so as to have systems which work to resolve the emerging issues.

This chapter takes a look at water management and governance-related issues in rural India at the district or local/regional levels.

Varying Water Uses

Roughly speaking, water is put to the following uses in the rural areas of our country:

(i) *Irrigation*, which includes crop husbandry, horticulture, silvipasture, afforestation.

(ii) *Domestic and drinking* water use in India's villages/hamlets and on farmers' fields.

(iii) *Drinking water* for the livestock.

(iv) Rural industries, including mining and agro-industries/agro-processing.

(v) *Environmental* uses for facilitating (a) *leaching* of soils and preventing soil salinity, etc., through *drainage,* (b) *disposal* of *wastes* sewerage, effluents, etc., and (c) *preservation of wildlife* sanctuaries, national parks and *wetlands.*

Existing Institutions

There is a multiplicity of institutions and agencies at the district, equivalent, or lower levels, which deal with the management of provision of water for the above purposes.

There are the district level representatives of the second tier of India's federalism, that is, District Level Officers (DLOs) of State Government departments and agencies including from the following departments: (a) Water Resources & Irrigation, (b) Command Area Development (which may either be separate or part of the Water Resources Department), (c) Agriculture,

(d) Animal Husbandry & Livestock, (e) Forestry, (f) Public Health Engineering Department (PHED) or Rural Water Supply, (g) Ground Water, (h) Rural Development (which may be merged with or function independently, of the Panchayati Raj Department), (i) Industries, and (j) Environment.

The District Collector/Deputy Commissioner/District Magistrate and the District Administration, the DRDA (District Rural Development Agency), and staff of special project organizations set up by various State Government departments as well as representatives of public companies like a State Industrial Development Corporation would also fall in this category of representatives of the second tier.

The third tier which comprises all three levels of Panchayati Raj Institutions (PRIs), i.e, Zila Parishad, Panchayat Samiti or Gram Panchayat, and even the Gram Sabha, are also involved in some of the water supply and management tasks.

Non-governmental organizations (NGOs), whether publically or privately sponsored, have also been involved at many places in water management, and the proponents of Panchayati Raj refer to them as "parallel bodies" with which the PRIs may need to compete.

Then, there are community-based organizations (CBOs) and user groups like Water Users Associations (WUAs), Watershed Committees, Joint Forestry Management Committees, Locality Residents' Committees for drinking water, etc., which are also important from the point of view of managing water in rural areas. This writer views such bodies as a sort of fourth tier in India's federal structure - a tier which is somewhat informal and which presently does not enjoy constitutional status.

Now to turn to the various uses of water in rural areas and the attendant water management issues.

Canal Irrigation

Let us start with irrigation for meeting mankind's food and fibre needs. The Irrigation Department functionaries are today charged with constructing dams and canals and ensuring the operation & maintenance (O&M) of the canals. However, neither by academic background, nor by on-the-job experience, are they attuned to interacting on a sustained basis with the farmer water users. Further, the concerns of those who are to manage the headworks, and the main canal system by nature need to be very different from those dealing with on farm water application and rotational running of minor canals so as to reach water in desired quantities and at the desired time at the canal outlets.

A number of differing institutional models have been tried out to deal with the problem mentioned above. Either the Collector and district administration, or the panchayat bodies, or extension staff of the agriculture department, or a multidisciplinary Command Area Development (CAD) project agency, or Water Users Associations (WUAs) have been used to provide the interface between main system management (normally done by engineers) and the farmer water users at the level of the minor canal and its outlets.

Where the Collector/district administration is made use of, it is normally once in a crop season when a consultation is held in a water regulation related committee to assess water available for the particular season and to determine when water is to be let into the canals and the canal running programme for the entire crop season. The Collector is too busy with his various multifarious duties and also does not have the technical advice to go into more detail or to ensure sustained consultations.

The PRIs also have diverse responsibilities., water management being but a small part of their mandate. Also panchayat boundaries rarely follow the topography-influenced irrigation command boundaries (a situation also common with regard to district or sub-division boundaries). Hence, panchayats can rarely intervene in an effective manner in canal water management issues.

The agriculture extension staff of the agriculture department of the State Government does have a better idea than most other government staff of the requirements and needs of the farmer water users, but they do not have the authority to sufficiently influence the irrigation department functionaries and have not always been able to win the trust of the farmer water users in each and every irrigation command. They are vital for dissemination to the farmers not only of agriculture extension messages, but also of irrigation extension messages related to optimal operation, maintenance and rotational running of canals, to improved on farm water application, and to water conservation and water management.

However, for the vital task of an interface between main system management (irrigation engineers) and on-field water application and management of minor canals (farmer water users) they need another agency to be involved which they could then assist in this task.

A few, but not all, CAD projects have an integrated and interdisciplinary project organization, where staff from various relevant government departments is placed in the integrated CAD project organization. Thus, irrigation, agriculture (including agriculture extension), groundwater, PHED and even afforestation and animal husbandry departments staff may all be placed on deputation in the same CAD project hierarchy under a common administrator. (It may be added that the CAD approach, ever since its introduction in the mid-1970s, has emphasized on an integrated, and multidisciplinary project approach for CAD but in practice only a few CAD projects are integrated, and multidisciplinary). Such integrated, interdisciplinary and inter-departmental CAD project set-ups can undertake many water management related tasks and also can act as an effective interface with the farmer water users. However, such integrated CAD entitles may only prove cost-effective for the larger irrigation CAD projects which may be cutting across many districts. Not every small irrigation project can have such an elaborate integrated CAD organizational set-up.

That brings us to Water Users Associations (WUAs) and the concept of Participatory Irrigation Management (PIM). The PIM concept envisages the possibility of WUAs taking over and totally managing entire irrigation systems not only with regard to functions normally performed by the State Government's Irrigation Department but also with regard to other activities related to irrigated agriculture in all its dimensions. In India, however, PIM, though backed by legislation in some states (starting with the Andhra Pradesh Act of 1997), has by and large been limited to the WUAs undertaking O&M of minor canals and (only in some cases) taking on the task of collection of irrigation dues (a task otherwise performed by Patwaris of either the land revenue administration, or irrigation department depending upon the different systems prevailing in different states of the Indian Union). For O&M of minor canals the WUAs have to depend upon irrigation department staff for governmental funding and for technical support. Since the rest of the irrigation system above the canal minor still continues with the irrigation department, the WUAs also need to depend upon the departmental staff for supply of the water. For agriculture and irrigation extension support the WUA has to work with the agriculture department staff at district and lower levels.

PIM – with WUAs taking over various irrigation, agriculture and water management tasks and associating with the Irrigation/Command Area Development/Water Resources Department in a joint partnership at managing entire irrigation systems (including their drainage needs) – has great potential. Even today, many WUAs have contributed significantly to improved water management with the area under irrigation and productivity going up with less quantities of water being used, or with water being supplied as per their equitable shares in both head and tail reaches, or with canals being repaired and maintained in a cost-effective manner, with water being supplied at critical times and in the desired doses instead of being allowed to flow as per preset schedules rather than as per the dynamic situation of the changing requirements, and with water indiscipline in canals being checked. On the other hand, there are also many WUAs which have remained ineffective or dormant.

What is required is that the State Government creates a facilitating environment, whereby the government departments reorient their working and develop genuine working relationships and partnerships with the WUAs, who (through training and orientation programmes and governmental support including a system of incentives and disincentives) should have their capacities increased. If farmer water users know that their WUAs would be given the latitude to function properly and would receive proper back-up and support from the government agencies at the project/local levels, then, they would be willing to pay more for the better O&M and improved water-related services.

There is another major problem related to management of canals and that is one of over-optimistic delineation of the commandable areas in each project which have been fixed unrealistically larger in area than that which can actually be commanded or managed. For this the refixing of commands of old projects and realistic planning of new irrigation projects is the only solution.

Rainfed Areas and Watersheds

Now to turn to the rainfed areas where rainwater harvesting and watershed development and management is the key. While watershed works have been undertaken at least since the mid-1970s, the Government of India had compiled and issued guidelines in 1994 which underwent revisions in 2000, 2001 and 2003. This writer feels that the 2000/2001 guidelines were more practical. These "common guidelines" envisaged that the watershed development programme be implemented through the Zila Parishad/DRDA. A District Watershed Development Committee (DWDC), with multisource membership of officials and non-officials, is to ensure coordination. Watershed projects are to be implemented by Watershed Committees (WCs), under the overall supervision and guidance of Project Implementation Agencies (PIAs), which are normally to be assigned 10-12 watershed projects covering an area of 5,000-6,000 hectares. The PIAs are to be preferably selected from amongst the PRIs, or government departments/agencies or a reputed NGO.

There are four-member Watershed Development Teams (WDT) in each district for providing the necessary technical inputs, which comprise of a member each from the disciplines of forestry/plant science, animal sciences, civil/agricultural engineering and social sciences.

The following CBOs are also associated with the work of watershed development/management

1. *Self-Help Groups (SHGs):* Constituted in the watershed area with the help of the WDT, these are homogenous groups having common identity who are dependent on the watershed but not owning land in the watershed areas, such as agricultural labourers, shepherds, etc.
2. *User Groups (UGs):* These groups consist of persons most likely to derive direct benefits from particular watershed work and holding land within the watershed areas. The UGs are to eventually take over the operation and maintenance of completed community works.
3. *Watershed Associations (WAs):* Comprising of members of the community directly or indirectly dependent on a watershed area organized into a WA, which should be a registered

society as per Registration of Societies Act, 1860. The WA is to meet at least twice a year to improve, monitor and review the watershed development plan, nominate members of the Watershed Committee, etc. Wherever a watershed is coterminous with a Village Panchayat, the Gram Sabha of the Gram Panchayat concerned could be designated as the Watershed Association.

4. *Watershed Committee (WC):* It is to consist of members nominated by WA from amongst the user groups, self-help groups, Gram Panchayat and a member of the Watershed Development Team for carrying out the day-to-day activities of the Watershed Development Project. Each WC to have a secretary to be a paid on an honorarium basis by the Watershed Association. If the watershed boundaries happen to be coterminous with those of a Village Panchayat, then the Secretary of the Gram Panchayat could also be Secretary to the WA as well as to the WC.

As will be seen, the participatory aspects and the need to ensure local managerial inputs for land-water management in watershed development work have been catered to through the institutional mechanisms described above. For the technical inputs required for watershed development/management, a Watershed Directorate/Commissionerate has been set up at the State Government level with its district or regional units being multidisciplinary and being manned by staff primarily from the agriculture department but also from animal husbandry, forest and rural development departments. Unfortunately, the Watershed Directorate/Commissionerate has recently been shifted from the control of the agriculture department to that of the rural development department which has raised the danger of its technical focus being reduced.

Rainwater, harvesting efforts should normally be undertaken as part of watershed development/management with the water harvesting structures and practices being part of the comprehensive watershed area plan. Such efforts would then get jointly undertaken by the State Government's agriculture, watershed, groundwater, and panchayats departments along with PRIs, CBOs user groups, etc. However, rainwater harvesting is also done without adopting the watershed area planning approach. This should either be discouraged and the water shed area approach insisted upon. Alternatively, the panchayat bodies should be permitted to undertake such work only after they have cleared with representatives of irrigation, agriculture, and groundwater departments that the possible unintended consequences of the rainwater harvesting effort would not outweigh its possible benefits.

There is an issue of the upstream/downstream impacts of the watershed or canal irrigation and CAD efforts which can be felt outside the watershed or irrigation command boundaries which this author proposes to discuss slightly later in this chapter.

Groundwater-based Minor Irrigation

As regards groundwater use for irrigation, the groundwater department primarily undertakes survey of changes in quantity and quality of the groundwater and groundwater levels, or the development of some groundwater resources. The groundwater department staff is neither mandated to, nor is it empowered for, or capable of, regulating groundwater use. Nor are the panchayat bodies, or the district administration able to undertake such regulation. Decisions about digging of wells, installation of pumpsets and the amounts of groundwater to be drawn depend upon countless individual farmers each of whom makes his own decision. Model Groundwater Regulation Acts have been drafted but not enacted. Indirect efforts like the stoppage of loaning for groundwater related minor irrigation (wells and pumpsets, etc.) in the identified dark zones (where overexploitation of available groundwater is already being done) have failed to have the hoped for impacts; In fact, with the spread of electrification and the provision of subsidized and flat rate electricity for well irrigation, the exploitation/overexploitation of groundwater has gone up in leaps and bounds. This is a sphere of rural ground water management which requires much more thought and creative action.

Drinking Water and Water for Domestic Use

Now to turn to drinking water and water for domestic use in the rural areas. Water may be supplied in the villages through hand pumps, community wells (traditional source schemes or TSS), tube wells, or from canals, or special drinking water pipelines carrying water from long distances (and, in times of drought, even by rail or road tankers).

Handpump installation, revival of community wells for the TSS schemes and construction of tubewells for village water supply is normally done by staff of the Public Health Engineering Department (PHED) in consultation with the Panchayat Samiti or Gram Panchayat. (For drinking water the quality of water and its being potable is equally, if not more, important than its availability in the required quantities.) Efforts are made to get the Gram Panchayat (which often instead prefers that PHED do the work and the panchayat body get to issue supervisory instructions without being held responsible for the results) to take over and operate and manage the TSS or tubewell-based schemes as also to take over responsibility for minor repairs, with PHED periodically launching major repair campaigns. For minor repairs the PHED arranges for training of personnel, whom it is expected that the Panchayat Samiti would use either on a task basis or fixed amount contract basis to get the repairs carried out.

Where water has to be brought into villages from outside sources (which may be a few kilometres away or even hundreds of kilometres away), then, the PHED has to construct the pipelines and to arrange for transportation of water from the source (a dam a large canal or a battery of tubewells) to the village. The PHED is also expected to develop the sources. However, since drinking water has the first priority amongst all water uses as per the national and state water policies, PHED often gets to take over part, or the entire share of water from the irrigation dams or canals, much to the chagrin of the irrigation department staff.

Drinking and domestic water supply for the villages is so vital that the District Collector and district administration play a much more proactive role in coordinating and monitoring this sphere as compared to the other water uses and also as compared to the work of many other departments. Of course, in times of drought (and at times even otherwise) recourse has to be taken to tankers to supply water to villages and in such cases, it is the District Collector (who also coordinates all drought/famine relief works in the area), who has a major role to play getting the work done through the PHED staff.

Farmers drawing water for irrigation from canals are expected to use part of it for domestic and drinking water purposes, especially if they live on their fields. Similar is the case if they have well-irrigated fields on which they live. In fact, in the sphere of rural water supply, all focus used to be on supply at one or two common points in the main village and it is only in recent years that attention has started shifting to hamlets or *dhanis.* Piped water supply is also not a common concept in villages, unlike in towns and cities, although theoretically nothing stops a panchayat body from arranging for the same.

Water for Industries in Rural Areas

For rural industries, large industrialists, including those involved in mining are expected to arrange their own water supply by developing wells or small dams with the permission and technical support of the government. Similarly, for industrial estates in rural areas, the State Industrial Development and Investment Corporation, which develops such estates, is expected to develop its own sources and to maintain the water supply within the estate. Only in some rural industrial estates situated near village or town habitations may PHED help out in the water supply. (The practice of reserving a quantum of the water in irrigation canals for industrial purposes, as has been done in the Indira Gandhi Canal Project, is not yet common.)

Environmental Requirements of Water

Irrigation drainage is totally within the sphere of the irrigation or CAD department. As regards other environmental uses of water, the Environment Department, whose main organ is the Pollution Control Board, is thus involved almost entirely in regulation rather than in developmental work. It prescribes norms and, particularly in the case of industry and mining, takes steps to see that they are enforced. It does not itself get involved in arranging for the disposal of effluents, or wastes. Similarly, the groundwater department has certain capabilities for monitoring water quality of both surface and groundwater but does not undertake works aimed at improving the environment, or the quality of water and the land. Thus, sufficient mechanisms do not exist for improving the rural land and water environment as different from protecting the environment. Therefore, this is a water use in rural areas regarding which a lot more needs to be done than is being done now.

Preventing Unintended Consequences and Ensuring Coordination of Efforts

The question of unintended consequences (which can be both positive and negative) of works within a watershed, or which form part of CAD, has been alluded to earlier in this chapter. Some such consequences can occur outside the boundary of the particular watershed or CAD project. Cases are known where for example anicuts constructed as rainwater harvesting devices having reduced surface water flows in downstream rivulets and streams which in turn have effected irrigation or drinking water availability in proximate areas; or, where canal irrigation in certain areas (where the farmers were not particular about efficient on-field water application or about canal rotation being managed so that canal flows are consciously restricted in periods of low water demand) has resulted in water logging of areas outside the command area. Other similar examples could also be quoted.

One lesson from such cases is that canal irrigated areas, rainfed areas and ground water areas need to be viewed as one continuum and to be planned for in an integrated manner. This would imply the need for intensive coordinational interaction amongst all the concerned government departments mentioned above – as also for mechanisms for the concerned departments to obtain relevant planning and coordinational inputs from the CBOs and user groups as also from the Panchayati Raj bodies.

There have been recommendations for the creation of a 'super' Water Resource Planning Department at state level which would concentrate only on analysis, and planning and whose clearance may be required by all other related departments before they can undertake any activity which impinges on water. However, a system of a 'super' department having authority over other secretarial or administrative departments may not prove workable in the Indian context.

A recent Government of Rajasthan innovation whereby groups of two or three related secretarial departments headed by secretaries have been brought under the single umbrella of a

Principal Secretary or Additional Chief Secretary has proved to be conducive at the top policy level. However, the problem identified by us calls for much more intensive coordination at the project formulation and project implementation levels involving detailed analysis of local (and regional) field conditions at various places spread throughout the state. It also assumes that detailed data is available or can be collected and analyzed in an inter-disciplinary and multi-departmental manner at the state as well as lower levels. The division level may be too far removed from the field to be the effective level. Even the district level may not fully suffice particularly as district boundaries are based upon historical or administrative factors while command area, or drainage catchment, or watershed boundaries are based upon topographical, hydrological and agro-climatic factors and thus they may not coincide at all with the administrative boundaries of either the district, the sub-division or the tehsil, or even of the Panchayat Samiti or the Gram Panchayat.

An ideal, perhaps even utopian, situation would be having a system covering the entire state of authorities like the few existing integrated, interdisciplinary CAD project authorities (which both plan and implement water-related activities) with staff from all departments concerned with water management and area development being on deputation to such authorities. The territorial jurisdiction of such bodies would need to be determined on topographical, hydrological and agro-climatic considerations and all water-related activities should be kept reserved for this authority to undertake or to permit other agencies to undertake after due examination of the likely intended and unintended consequences. However, we cannot now reorganize each state into such authorities. Nor would it be easy to deal with the administrative problems associated with operating such agencies.

What may be more practical is to make use of such integrated authorities where they already exist – and where they do not exist, there a core technical cell (having links with a similarly created multidisciplinary cell at the state level having access to all relevant data of all concerned government departments) should be made operational at the district level and it should service an interdepartmental water resources committee under the District Collector, which could also interact with a parallel committee of PRIs and

CBOs and water user groups. The Collector and committee at the district level should draw heavily from the data and technical competence available in the core technical cell. Clearance for the initiation of any water-related activities in the district should need to be taken from the inter-departmental water resources committee at the district level. The committee would need to adjudicate between alternative uses of water as also between demands of different sub-areas while examining the techno-economic and social as well as environmental soundness of proposed activities and practices proposed to be followed by each activity, programme project or sub-project. However, it needs to be cautioned that making even such a committee system work would not be easy. Also, there is the fear that all water-related efforts would get delayed as such a committee may find it difficult to grant clearances quickly and thus may become a stumbling block.

Growing Competition and Emerging Conflicts over Water

There are many who feel that the Collector is too busy with the countless activities assigned to him and that the Collector, being a generalist administrator, may not be able to do full justice to the technical issues coming up before the committee as advocated in the preceding section of this chapter. Some such people would advocate the placing of a water management expert to head the committee. However, the multifarious tasks being performed by the Collector enable him to take a broad interdisciplinary view of issues as compared to a technocrat who often tends to get limited by the monodisciplinary nature of both his initial academic background and his experiences within his parent department/ discipline.

However, there is another reason why the Collector should be kept as the king pin of the proposed committee system. Competition, agitations and local conflicts over water have been gradually growing. Villagers forcefully want to retain water which rains on their village catchments from flowing downstream even if it means the drying up of rivers or rivulets which have been flowing since times immemorial. Protests have been made in an effort to stop established systems of water being supplied to nearby sanctuaries, or towns, on the ground that the water is now required by the villagers for irrigation. Livestock breeders on their annual summer migration have at times been denied water (and pasturage) at places they have traditionally relied upon. Digging of wells to supply rural industrial estates is often resisted on the grounds that the proposed water drawals would lower water tables and thus adversely effect well based irrigation in the local area. Such examples are increasing and they very often assume law and order dimensions.

It is the District Collector-cum-District Magistrate with his foot in both regulatory (including law and order) and in developmental administration who is best suited to tackle such law and order problems. In fact, problems can often be prevented by advance planning and pre-emptive and pro-active decision-making. The committee system as suggested in the proceeding section can also be used for this purpose in case the Collector heads the committee and is the key coordinator – of course with concerned DLOs, panchayat bodies, CBOs and the core technical cell all having a suitable say, or voice, in the matter.

The system suggested would need to be worked out in more detail along with careful designing of the committee and technical cell structures as well as the processes for interaction and consultation with all concerned. It would then need to be introduced in such a manner that the transitional disruption and cost is minimal and refinements in the system can be undertaken and new innovations tried out as per experience gained in the field.

Only thus could a new paradigm of governance and improved management of the rural water sector emerge – a paradigm of water governance that works, particularly at the local and regional levels.

Chapter 5

Water Management

An Environmental Perspective

Brij Gopal

AMONG all other substances on the earth, water is unique. Water is life; it sustains life and there would be no life without water. For humans it is not simply a biological necessity; the human civilizations grew around water: human cultures and all human endeavours depend on water. Chemically, water molecule has so strong bonds that it changes its form from gas to liquid to solid but is difficult to split. Water forms the basis of bonds between all living organisms and non-living matter on the earth. More than two-thirds of the earth is covered with water – with oceans deeper than the highest mountains, polar ice caps and the snow. Yet, humans and the majority of organisms can use only a small fraction, less than 0.01 per cent of the total water, that is fresh, and which flows in the rivers, lies in the lakes or below ground aquifers. This freshwater is highly variable, spatially and temporally, both in amounts and quality.

For several thousand years, humans have used water for growing food but during the past two centuries or so of industrial revolution, human activities have greatly diversified causing increasingly greater demands on water. At the same time, the exponential increase in human population, coupled with urbanization, has increased the water requirements manifold. Multiple competing demands on a critical resource in limited supply naturally calls for management, of which various facets are discussed in this volume.

Most management discussions, however, focus only on the amounts of water, readily forgetting a few basic facts. First, water is different from all other resources. It is both a renewable resource and a non-renewable resource. It is renewable because it cycles continuously between the ocean (hydrosphere), air (atmosphere), land (geosphere) and the organisms (biosphere) (see Figure 5.1). But it is non-renewable because the amounts cannot be increased. Water is not fixed like land that can be divided into portions and allocated for different uses. Unlike land, water can be carried from one place to another and can be stored for use whenever required.

Figure 5.1
The Hydrological Cycle

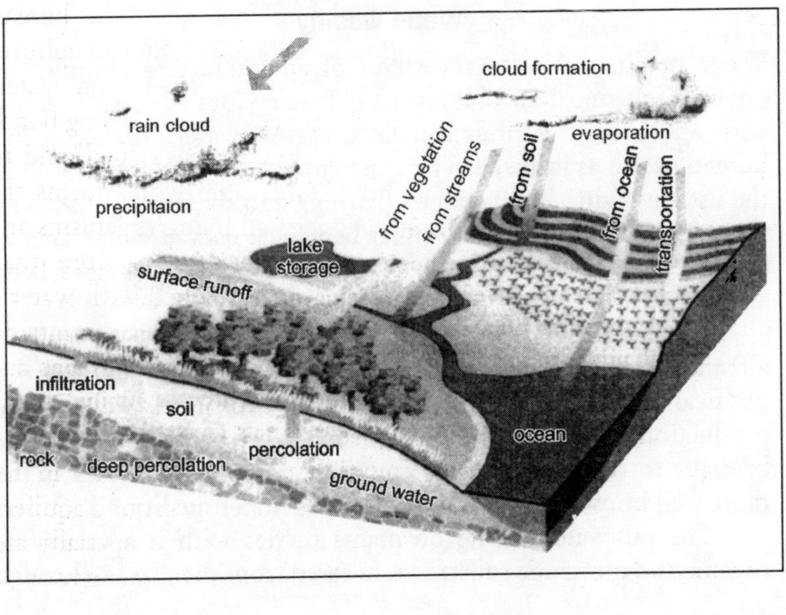

However, each and every use of water changes its quality and every other human activity howsoever far away from the water impinges on its quality. Finally, the quality governs its use. Water is valued for its quality far more than for the amount. A few drops of water from near the source of river Ganga are even today more valuable for their purity, albeit symbolically, than the water in the huge oceans that can be used for neither irrigation nor domestic supplies. Second, it is a common knowledge that flowing water remains cleaner whereas the stagnancy degrades its quality. Finally, and most important, water is not for humans alone; myriads of organism live and depend entirely upon water as their habitat. These organisms are a must to keep the water clean and clear as they use up the nutrients and feed on organic matter. Bacteriophages are believed to be responsible for the purity of Ganga water and its keeping properties.

In the following pages, therefore, I dwell upon the need for emphasis on quality in the management of water resources, and plead on behalf of the millions of 'speechless' stakeholders – plants, animals and micro-organisms – who not only live in water and hence need water but whose survival and welfare is critical to the human welfare.

Water Quality

Water quality refers to the status of an interacting complex of physical, chemical and biological characteristics of a water body with reference to various beneficial uses of water. Traditionally, human health as influenced by consumptive use of water as well as the use involving body contact (bathing and swimming) has been the primary concern. Many other beneficial uses of water such as that for fisheries, agriculture, industry and recreation have been acknowledged.

It is determined by a host of processes that occur on land, in air and water and within the biological communities. Various physical and chemical processes in the atmosphere result in the production of chemical substances such as nitrogen oxides that are brought to the earth during rain and snowfall. Acid rain is the most well-known phenomenon affecting water quality.

The rainwater falling on plants carries with it a variety of organic and inorganic substances released from the plants (through leaves and bark) as secondary metabolites, besides the particulates deposited on the plant surface from the atmosphere. Both organic and inorganic particulate and dissolved substances are carried by the runoff over ground. The nature and amount of these substances depend upon geology, physiography, climate, soil characteristics, vegetal cover, fauna, and human activities. The infiltration of water through soil removes soluble substances to the groundwater, and also influences several chemical reactions within the soil that also affect the chemical nature of the groundwater. Sub-surface flows moving laterally affect the surface waters. The biota (plants, animals and micro-organisms) exert far more important and greater influence on both the amounts and quality of water.

Regulation of Water Quality

Plants play an important role in determining the water quality, by absorbing nutrients and releasing them during decomposition, and contributing organic matter. Many saline plants exude salts from their leaves. Many plants take up, and accumulate in different organs, varying amounts of toxic substances (such as heavy metals and pesticides). Some plants enhance or reduce the loss of water through evapotranspiration. Further, they aid and abet siltation. Some animals such as rodents and otters make burrows or tunnels on lakeshores and riverbanks, which may result in increased erosion and flooding. Grazing and wallowing by larger animals such as buffaloes and rhinos also affects water quality.

Biological processes such as photosynthesis by submerged plants cause precipitation of calcium and phosphorus from the water column. Transport of oxygen from leaves to the roots, and its diffusion into the soil, causes the precipitation of iron and manganese on the root surfaces. The organic matter added by the plants to the water body may help in the reduction of nitrogen and sulphur through reduction reactions (e.g., denitrification). Animals help remove the organic matter which they feed upon. Many fishes are therefore grown even in sewage as a simple cost-effective method of wastewater treatment.

The micro-organisms are even more significant to water quality. Whereas many of them (both bacteria and blue-green algae) are undesirable, planktonic algae utilize nutrients and contribute oxygen to the water body. The blue green algae enrich water with nitrogen from the atmosphere. Various bacteria and fungi help in the decomposition of organic matter, and transform various nutrients and toxic substances.

Various ecological processes assume their greatest significance at the interface between land and water, i.e., in the transitional zones such as shallow peripheral zones of lakes and the river floodplains better known as wetlands. These areas are not only highly productive and provide valuable forage to a range of grazing animals, they help improve the quality of water in the rivers and lakes by preventing flow of silt, nutrients and toxic substances (see Loucks, 1990; Wetzel, 1990; Naiman *et al.*, 1993; Gopal, 1994).

Causes of Water Quality Degradation

The most common water quality problems include: oxygen depletion and consequent changes due to discharge of organic wastes, eutrophication from the abundance of nutrients, health hazards from pathogenic organisms and a variety of persistent toxic substances including heavy metals, high turbidity due to large amounts of suspended matter, salinization due to high concentrations of ions such as calcium, sodium, chloride and sulphate, and acidification from atmospheric deposition. All streams have a limited capacity for self-purification as the wastes are metabolized by bacteria, fungi, invertebrates and many other animals (Patrick, 1975). The assimilative capacity depends upon the rivers' flow regime (volume, velocity, duration and timing) and interaction with their floodplains. Water quality problems therefore arise when input of wastes exceeds the assimilative capacity.

All kinds of anthropogenic activities within the water body or on land or in the atmosphere impinge directly or indirectly upon water quality. The gaseous and particulate products of burning of fossil fuels released into the atmosphere are returned to the water bodies through precipitation and runoff. Clearing of vegetation, overgrazing by domestic animals, mining., soil excavation and agriculture affect water quality through erosion and loss of nutrients with the runoff. Direct discharge of domestic and industrial effluents affects the biota directly and disrupts the nutrient cycles. Alterations in aquatic habitats directly (for example, by land fills and construction of embankments) or indirectly (by regulating river flows or altering the runoff) cause severe changes in biological communities and ecological processes that consequently degrade water quality.

Management of Water Quality

Thus, quality is not a simple matter of the presence of a few visible or dissolved 'impurties' in water. It is affected by the events and human activities not only in water or along the lakeshores or river-banks but far beyond the shoreline in the entire catchments. It affects not just the humans but all other organisms in the water bodies. Thus, water quality is concerned with the total "ecological integrity" of surface waters (see also Lijklema, 1995).

As said earlier, all uses of water are directly related to its quality and therefore, water management has to include all issues concerned with quality. Traditionally, water quality management has required treatment of wastewater to reduce the concentration of organic matter and nutrients to levels that can be readily assimilated by the streams and rivers. Both organic and other solid wastes were prevented from entering the water bodies in large amounts. All methods of secondary treatment of wastewater utilize the services of micro-organisms in transforming dissolved organic matter and nutrients to particulate matter (sludge) by subsidizing their energy and oxygen requirements. Tertiary treatment in some countries utilizes the services of aquatic or terrestrial plants in order to further reduce the nutrient concentrations and to some extent, toxic substances.

However, activities such as the landfill of shallow marginal areas of rivers and lakes, exclusion of floodplains by embankments, and elimination or large reduction of flow by storage and withdrawal of water (by constructing dams and barrages), almost totally destroy the natural biota and their ecological functions. On one hand, the managers of water resources emphasizing upon the amounts alone, focus on the creation of new water storages – tanks and reservoirs, on the other this very action strangulates the existing water bodies. The removal of floodplains or their water supply directly kills the aquifers which depend for their recharge upon the annual floods. The current water management practices that seek to increase the productivity of the land and support the livelihood of the farmers – many among them neo-rich cash crop cultivators, in fact eliminate the productivity of the natural water bodies, especially the riverine fisheries, and strike at the very source of subsistence of millions of poor people who traditionally depend entirely upon the rivers and lakes for their livelihood. Also lost are hundred of organisms which cannot raise their voice or even move to another 'habitat' while the policy – and decision-makers claim their support for the conservation of biodiversity!

Management of water quality requires several actions. First and foremost is the immediate need to prevent generation of wastewater by its efficient utilization and recycling. Simultaneously, the consumption of water must be reduced in all sectors – domestic, agriculture and industry – by employing appropriate methods. However, it must be appreciated that the increasing demand of water to meet the needs of a growing human population will still continue to generate enough waste and wastewater that cannot be treated conventionally and assimilated by the degraded, depleted remnants of inland aquatic ecosystems. Therefore, there is a more urgent need for restoring the natural ecosystems and let nature function to treat our wastes without provision for external subsidies. Technologies alone cannot provide solutions for ecological problems. We must respect and restore nature's ecological integrity as much as possible. Water management must consider the water bodies as 'ecosystems' interacting with their catchments on a river basin wide scale. The water managers must leave their myopic vision of water as a tradeable commodity; instead they must learn to visualize and comprehend the entire landscapes.

■ References

Gopal, B., 1994. Role of ecotones in conservation and management of tropical inland waters, *Mitt. Int. Verein. Theor. Angew. Limnol*, 24: 17-25.

Lijklema, Lambertus, 1995. Water Quality Standards: Sense and Nonsense, *Water Science and Technology*, 31(8): 321-327.

Loucks, O.L., 1990. Land-water interactions, pp. 243-258. In: Patten, B.C. *et al.*, (eds.), *Wetlands and Shallow Continental Water Bodies,* Vol. 1, SPB Academic Publishing, The Hague.

Naiman R.J., Decamps, H. & Pollock, M., 1993. The role of riparian corridors in maintaining regional biodiversity, *Ecological Applications,* 3: 209-212.

Patrick, Ruth, 1975. Some thoughts concerning correct management of water quality, pp. 16-18. In W. Whipple, Jr (ed.), *Urbanisation and Water Quality Control,* American Water Resources Association, Minneapolis, MN, USA.

Wetzel, R.G., 1990. Land-water interfaces: metabolic and limnological regulators. *Verh. Int. Verein. Limnol,* 24: 6-24.

Chapter 6

Community Water Management

G.S. Narwani

INDIA has been a welfare state committed to democracy, development and decentralization of powers. People's participation has emerged as a key developmental strategy and it has been given a concrete shape in form of Panchayati Raj Institutions (PRIs). People's participation in formulation and implementation of rural development programmes has unfortunately not received the proper attention and importance it deserved.

Various programmes for development of rural areas as well as welfare measures for poor people have been implemented for more than 50 years but even then the improvement in their economic and social conditions is much below the accepted level. They must be motivated to play more active participatory role in the entire process.

The natural resources, which are a boon of nature to mankind, had been controlled by the state government departments of revenue, forest, irrigation, fisheries, public health engineering, etc., for earning revenues for the state. They were badly managed in the past. Poor people were denied its benefits since mostly they were monopolized by few landlords, big farmers, contractors, industrialists and influential sections of society. Hence, now, it is high time to vest back these natural resources to the people themselves through Gram Sabha or user committees for proper management, preservation and development to the optimum possible extent and to the best interest of the community as a whole including poor and marginalized classes.

Moreover, the resources on which people subsist for their living cannot be treated as property, that the people have an inherent natural right over the same as a means of livelihood which is superior to state's claim. The idea of inalienable right of the people to use resources for their livelihood, the questioning of displacement for development which essentially means alternative use of resources, the rights of self-governance lose much of their significance without command over the habitat. Concentration of wealth and economic power goes against the constitutional provision of Article 39 which enjoins the state the responsibility to ensure that the community resources are so used as to subserve the common good. Moreover, it is now asserted that the command or ownership of the community are non-negotiable. A person can best assume the role of a manager but the resource vests in the community.

The National Water Policy, 1987 advocated 'participation' of farmers in management of irrigation systems. It said that "farmers should be involved progressively in various aspects of management of irrigation systems, particularly in water distribution and collection of water rates". This was encouraged by the Government of India when it was laid down in Eighth Five-Year Plan document, 1992-97 that "the farmers can take over the maintenance and operation of the system themselves".

The then President Dr K.R. Narayanan, while inaugurating conference organized by the Centre for Science and Environment on 'Potential of Water Harvesting' at New Delhi on 3 October 1998 warned the nation: "This (water) universal asset is being fast depleted. Leading experts on water resources have been warning that the world is heading towards a 'Water Shock' which may dwarf the oil crisis. Some 80 countries and 40 per cent of the world's population are already experiencing 'Water Stress'. Environmental degradation and population explosion are making the problem even more acute, with pollution poisoning the water that is available." Since recently 'Battle for Water' has started by the community against new 'Water Policy' of the government. Water is not a private property, but a common resource, a boon of nature. The right to conserve, use and manage water is fully vested with local community. This is the very basis of water democracy. During the past also, community management of tanks, tubewells, lift irrigation schemes, watershed projects, water harvesting structures, canal irrigation and drinking water programme have been in practice. Water which is a community resource, is being treated as a state monopoly and is given to people only as a 'gift'.

Need for the Community Involvement

Former President of India Dr K.R. Narayanan stressed upon people's participation emphasizing that "India today needs a people's movement to meet its water needs and to protect its water sources. We must get every community involved both in the provision of water and in the protection of water sources. Civil society has to play a critical role in encouraging equity and sustainability in the use of water." Increase in human and bovine population of the country has resulted not only in reduction of per capita availability of net area cultivated but also in substantial reduction in availability of common properties for grazing for the cattle and shortage of fuelwood for domestic purposes. There have also been increasing number of cases of garbling of common lands by influential people of the villages. In earlier years after independence, the pressure on lands was not so much because there was enough land available for people as well as cattle, no body thought of reclaiming waste lands or conserving flowing waters, but after increased demand for agriculture and diminishing water levels, need for management of natural resources has become imminent. The present status of management of forests, land and water is a big question mark since environment is degrading fast in spite of the best efforts of the government and biomass is almost disappearing not only from the pastures but also from reserved forests what to talk of unreserved forests. Hence, it is already high time that the community should come forward to protect common property resources to save it from the shackles of unscrupulous hands in the common interest. Community should come forward boldly for management of natural resources to get maximum and purposeful returns from land, water and forest.

Objectives of Community Water Management

Following are the major objectives of community water management:

(i) Right of water to the community and responsibility of the community to conserve water and equitably distribute among members.

(ii) Create awareness regarding associating people with rivers rather than joining of inter-state rivers.

(iii) Recharging ground water through sharing past experiences of benefits of water conservation projects executed by the community.

(iv) Get such water policy prepared at the government level in which community participation is ensured and water is treated as a joint property with the community.

(v) Right for use of water for those who conserve water.

(vi) Create awareness among people regarding water conservation and create community forum for water management.

Natural resources cannot be managed and preserved properly without involving the rural communities which derive their sustenance from the natural resources. Land, water and forest are really the source of livelihood of people in rural areas. These cannot be utilized for commercial purposes by the state. Community has the right to use for their subsistence. The aim of the land use policy is to provide supportive technological and institutional structures to enable these rural communities to manage the resources in an optimum manner

Community natural resource management system has existed in the historical past. New people-based initiatives have also emerged widely in recent years. The *Van Panchayats* of Kumaon, *Pani Panchayats* of Maharashtra, *Sukho Majri* experiment in Haryana and *Ralegaon Sidhi* of Maharashtra, Gopalpura in Alwar (Rajasthan), *Arabari* in West Bengal and JFM Committees of Orissa, *Chipko* in Himachal Pradesh, Vankar Cooperatives of

Gujarat are some of the well-known examples of participatory efforts of the people. In above cases, people put an end to free grazing, developed waste lands through water and soil conservation and revitalized agriculture, animal husbandry, horticulture, etc., and shared common benefits of tree growing.

Role of the Government

In such a historical process, the role of government should evolve from that of being a systems manager to that of being a service provider. Farmers should move from water recipients to water users and then as water co-managers along with irrigation department. But, this is not an easy process for agencies to accommodate beneficiaries and make space for so-called reserved domain of administration. There can be many ups and downs in performance and examples of success and failures may be there. But participation is essential in view of prevailing conditions specially when it is not possible for the state to maintain the irrigation as well as drinking water infrastructure due to financial constraints. This is true of not only India but other countries in the world also. It is now felt that one of the most important reason for these problems is lack of stakeholders' participation in water management.

The role of government institutions in the management of water resources is changing partly due to two trends:

1. Most governments around the world are confronted with the reality of decline in fiscal capabilities. They do not have adequate financial and personnel resources to continue to operate water management systems as directly as in the past.
2. With increase in literacy levels and general consciousness among cultivators and community, people have been capable enough to take care of mutual interest and manage community affairs to the best of their capability. At the most some orientation will build their capacity to manage things well.

In general, state governments should withdraw from doing things that farmers/villagers can handle

(i) to reduce the costs of organization and maintenance but also

(ii) to fine-tune O&M since users have more detailed knowledge of local conditions and have more stake and motivation to utilize available resources more efficiently.

Example of running *Pani Panchayats* in Maharashtra has illustrated how farmers made optimum utilization of irrigation waters and assured irrigation of at least two acres to each farmer in drought conditions.

This process should not be seen as irrigation or public health engineering department losing control over water and as giving up their responsibilities. Rather, control is exercised more indirectly but probably more effectively because government will now become 'wholesaler' of water rather than 'retailer'. A more participatory system of management should be able to save money and provide better service, if provision is made for accountable systems of organization.

Integrated Effort of Participatory Water Management

Participatory management of water may not be a panacea for all water management issues, but yet it can go a long way in ensuring efficient running of canals, optimum on-field water application and proper drainage of excess water from irrigation systems and farmers' fields. Such participatory management can work only if all the stakeholders – farmers, irrigation agency employees including engineers, government, etc. – feel that participatory management will further their individual and group or community interests, both singly and jointly. Farmers should feel an urge to come together in water user associations and to take over water management functions. Government employees should realize the need to change the ways. Training and orientation of both farmers as well as irrigation or other water agencies is needed for preparing all the stakeholders for success of participatory management.

Rajiv Gandhi Foundation Drinking Water Mission, along with donor, agencies also gave priority to community-based drinking water supply schemes. The 'Sector Reform' project was initiated in 55 districts of India including four districts of Alwar, Sikar, Rajsamand and Jaipur (Rajasthan) in the first instance some four years ago. Since 26th January 2003, it has been started in all the districts under the new nomenclature of *'Swajal Dhara'*. The scheme envisages 10 per cent contribution by the community and complete planning, execution and management of the water supply scheme by the rural community.

Thus, community management is becoming increasingly pertinent in the management of common property resources. Though ownership and management of these resources rested with the state for a long time, the government's failure to preserve common property resources together with their excessive exploitation for development has led to serious degradation of the environment with the attendant ecological repercussions. Under people's participation as an alternative, the beneficiaries are made responsive to the cost of preservation. Such a system makes development with preservation possible.

Real Objective

Community participation in water management should be genuine partnership between farmers and the government as a whole in irrigation and not merely as an instrument for facilitating government schemes and programmes. Solving fiscal problem is not the goal. Farmer's direct involvement in system management as a compliment/substitute to state's role should be the basic objective. It is not privatization which refers to transfer of ownership to the agency but direct and effective involvement of beneficiaries in decision-making process.

From Participation to Management

The idea that the farmers should 'participate' in irrigation management has grown in India since 1980s though formally adopted in National Water Policy, 1987 and Eighth Plan. Irrigation projects proved a disguised boon only. People's participation was fervently advocated by many sections of society. Bureaucrats saw possible benefits in probable reduction in operating costs, lightening their burden of maintenance, collection of water charges, etc. Farmer's perspective is satisfaction of his genuine need for irrigating his fields. He wants equity in distribution of water if it is scarce specially at the tail end. Water User Associations (WUAs) are to be used to develop them as a power base and to empower the community to take over management of water resources. It should come from below and not imposed from above if people have to manage and be really empowered to take over operation and maintenance of the system. Participation is only advisory where decision-making power rests with the agency and not with the users. Hence, it is 'management' and not 'participation' which can really deliver the goods. Farmers are the chief clients who can take initiative for irrigation assessment, planning, management, equitable water distribution, collection of water charges, maintenance, etc. Farmers are not only the beneficiaries but the managers of the product (water) which they purchase and for which they pay to the supplier and be accountable. There should be no paternalistic attitude.

The experience of various states indicates that 'kuhls' as managed by community in Himachal Pradesh, tubewells in West Bengal or Uttar Pradesh, tanks in Tamil Nadu, were of course successful. In pilot projects also, there was initial success but community management on sustainable basis is yet an exception. A climate is yet to be built for changing the mindset of irrigation officials. Community organizers have to create an initiative and build capacity among farmers to take over management or irrigation systems in times to come. Legal enactments alone as passed by different states may not be the way forward.

Pani Panchayats of Pune District

During 1972-73 the state of Maharashtra suffered one of the worst droughts. Purandar taluk of Pune district was not an exception but suffered acute shortage of water as well as food. An NGO – Gram Gaurav Parishad (GGP) – developed the concept of 'Pani Panchayats'. First such Pani Panchayat was formed for a patch of 16 hectares of almost degraded temple land situated on the hill slopes in Nayagaon village of Purandar taluk. Following five principles were decided:

(i) Only group irrigation schemes would be taken up on community basis.

(ii) Each member will be entitled to get water for half acre of irrigation with an upper ceiling of two-and-a-half acres for a household.

(iii) Water rights are not attached to the land. In case of sale of land, the new owner will not get water right. It was done to prevent the sale of land by poor persons due to economic compulsions.

(iv) Cost of the scheme would be shared between beneficiaries in the ratio of 2:4:4 among beneficiaries: GGP: Subsidy from the state. Beneficiaries will execute and operate the scheme.

(v) Sugarcane and banana crops are prohibited since they consume more water. It will benefit maximum number of members and maximum land.

There are about 60 Pani Panchayats in Maharashtra at present. Pani Panchayats have considerably raised the income and employment. Cropping pattern and crop intensity have changed favourably. Distribution of water is equitable. High water consuming crops are absent. Community need brought the villagers together and the experiment is running successfully.

Rejuvenation of Ruparel River in Thanaghazi, Alwar in Rajasthan

Rajendra Singh, Magsaysay Award winner, is now known throughout the country for 'National Water Awareness Campaign, 2002-03' for enlisting community awareness regarding water policy framed by the Government of India regarding engaging foreign companies on contract basis for supply of water. His NGO 'Tarun Bharat Sangh' (TBS) is a group of devoted workers for forest and water conservation incessantly working through the community in villages of tehsil Thanaghazi of district Alwar in Rajasthan since 1975. Along with water conservation, the community has also developed their local conventions for regulation of use of water. People take their own decisions at the community level now. Ruparel river dried down from December to July though its total catchment is 3,250 sq. km. Gujjars live in hills and rear cattle whereas Meena community is engaged in agriculture. Workers of TBS started organizing both women and men in the villages since 1975 for forest and water conservation. There was an acute famine from 1980 to 1985. There was no water in the villages. *Johad* (tank) was excavated by women of the village Malatolawas. During the first year, water remained for three months, during the second year it remained for 6-7 months and during the third year it lasted for the whole year. Workers of TBS were also engaged in *shramdan* with the women of the village. It prompted the villagers of other villages also to organize community effort. TBS also got financial assistance from IKKO in the beginning and from SIDA and INTERCOOPERATION later for water and forest conservation.

TBS conducted journeys on foot and organized community for *shramdan* during village meetings. They resisted even forest department and constructed check dams for water conservation in their villages and excavated tanks. Drinking water became available, water level in the wells rose high, migration stopped. People were happy. Though only four checkdams were constructed by villagers through local technical knowledge in 1987, in all 21 anecuts, 240 *johad/johadi* (big and small check dams), one dam, one fishery pond and contour bunding at more than 91 places were constructed by 1995 by the community under local guidance of community volunteers. It is exemplary case of community participation in the recent past.

All the success stories prove that community alone cannot achieve results unless some devoted social activists form a group of sincere workers who work incessantly in rural areas with missionary zeal. Motivation, guidance, training and support to community in form of technical and financial assistance combined with involvement of local community and help and assistance by concerned government departments can create such examples.

Tank Irrigation in Andhra Pradesh

There are about 74,000 tanks in Andhra Pradesh which irrigate 1.2 million hectares amounting to nearly 26 per cent of the area under irrigation. In all, 90 per cent tanks have a command of less than 40 hectares. Tanks are very useful in drought-affected areas of Telangana and Rayalseema regions. Traditionally, tanks provide drinking water and were constructed during 16th and 17th centuries by the then rulers. Tanks constructed by Vijay Nagar dynasty need special mention as after construction, the kings handed over subsequent management to the local community. Tanks are a classical case of community managed irrigation system in the country. It was tank water which, in fact, kept the villagers together. Though community management got a setback during British regime but even today people do involve in the tank management. Positive effects of efficient water management and reduced cost of irrigation is still visible. No negative effects have been highlighted in studies done so far.

Tank water flows from the sluice into the distribution chamber from which major supply channels take off. Earthen channels are maintained by farmers before letting water. Informal committees are formed by the farmers for cleaning and maintaining the channels which involve silt removal, uprooting vegetation and strengthening the bunds. Members are chosen by consensus. Generally, educated people are preferred. Each member should be a good cultivator and normally from major communities such as Reddis and Valmikis. Minor communities like Harijans, Chakala and Mangala do not find representation. The committee keeps a watch on water inflow into the tank and accordingly decide upon the time for opening the sluice. Before opening, the channels are cleaned under the supervision of the committee. Farmers committee appoints '*neerugantee*' to attend to water distribution beneath the sluice, who works under the supervision of the committees. He is paid both in kind and cash.

Regarding maintenance, all farmers are expected to take part. Those who do not attend, are charged Rs 25 per hectare and labourers are hired to do the cleaning. Thus, the responsibility of maintenance rests with the village communities, whereas administrative control of the tanks is with the irrigation department.

Tank Irrigation in Karnataka

At present, there are approximately 41,000 tanks in Karnataka. Tanks were constructed as early as in 12th century. Most of the tanks are managed by PWD. Farmers prefer to manage the affairs below the outlet.

Farmers organize an 'outlet committee' and prepare and implement the roster for the outlet for water distribution. Representatives from each outlet committees and local officials form a 'Tank Irrigation Committee'. Irrigation Management Committee will draw up the roster for distribution of water among the outlets. Outlet committee implements it below the outlet. PWD maintains main canal. Conflicts are resolved by farmers themselves in OC and IMC looking to nature of dispute. Communication between the farmers and departmental officers is very limited which needs to be developed for more efficient results. There is a need for more effective mechanism to ensure the adherence to the roster of water supplies providing access to all the farmers.

Watershed Projects

Towards the end of 1994, the Hanumanthaiya Committee appointed by the Government of India recommended:

(i) Water conservation and not soil conservation is the most important. Water holds key for development of rainfed areas. If water is conserved, soil would be conserved automatically.

(ii) It is also impossible to cover the problem areas in foreseeable future through the government machinery/set-up. Necessarily, the projects will have to be planned, implemented and maintained through active and willing participation of the people.

(iii) There will be community participation in soil and water management only when they perceive direct and tangible benefits for themselves and understand technology.

(iv) The full participation of the watershed committees may be facilitated when the programme is built up on their accumulated wisdom, skills and systems. The complicated technologies imposed from outside inhibit willing participation of the local people.

New approach of participatory natural resource management would enthuse the people, make the development programmes really useful for the local people, and make them come forward to welcome and participate in the great national venture towards sustainable management of the natural resources to produce abundance for our growing population without damaging the resource base. Basic changes envisage:

- Instead of government projects with people's participation, they should be people's projects with government participation. People's participation has to be ensured throughout the project cycle.
- Training will be given to farmers on participation, managing and working of groups of beneficiaries and technical matters related to crop production and soil and moisture conservation practices. Women farmers will have a high priority while selecting trainees and project will make full use of their experience wherever possible.

- It will ensure sustainable management of natural resources because the community watershed committee will not only plan and execute but also maintain the assets after completion of the project period.
- Beneficiaries would contribute at least 10 per cent of the project expenditure which will remain deposited as 'Development Fund' to be utilized for maintenance after the completion of the project over a period of five years.
- All the development expenditure is placed at the disposal of the local committee which prepares the account under the guidance of the WDT who verify its genuineness and committee submits it to the sanctioning authority.
- Indigenous technical knowledge of the community is to be utilized compulsorily rather than modern techniques. Local cheap material has to be used on priority.
- A high degree of flexibility and targeting of location specific problems in project planning and implementation is required.
- Priority is given to panchayats for implementation of the watershed projects,
- Women, SC, ST, panchayat members are also to be included in the Watershed Committee to ensure their participation and empower them through decision-making in development of the village.

There will be community participation in soil and water management only when they perceive direct and tangible benefits for themselves and understand technology. The full participation of the watershed committees may be facilitated when the programme is built up on their accumulated wisdom, skills and systems. Instead of government projects with people's participation, they should be people's projects with government participation. People's participation has to be ensured throughout the project cycle.

Best National Productivity Council Award Winning National Watersheds of Rajasthan

Watershed Development Projects and Wasteland Development Projects are executed by all the states and they are evaluated at the national level by a high powered committee. During following years, watersheds of Rajasthan state were adjudged as the best in

India and they were awarded National Productivity Council (NPC) best award at the national level.

Year	*Name of Watershed*	*Award*
1992-93	Losing (Udaipur)	Best watershed of the country for the year
1995-96	Kamli (Rajsamand)	Best watershed of the country for the year
1996-97	Balesaria (Bhilwara)	Best watershed of the country for the year
1998-99	Dumara (Ajmer)	Best watershed of the country for the year
1994-96	Kotri (Bhilwara)	Best Rajiv Gandhi Wasteland Development Project of the country for the year

Main Features of Community Management

- General awareness camps.
- Visit to successful watershed areas to arouse interest about utility of watershed development project.
- Entry point activities to associate people with the project.
- Formation of homogenous Self-help Groups (SHGs) of water users, animal breeders, shepherds, women, nursery group and such other beneficiaries having common interest in the project.
- Group meetings to identify group needs and prepare plan accordingly as per felt needs of the villagers.
- Participatory rural appraisal of local resources.
- Formation of Watershed Development Committee from among the beneficiaries.
- Preparing need-based plan.
- Entrusting responsibility for execution of the programmes among different groups by the WDT.
- Actual execution in the field.
- Monitoring by members of the committee and collecting contribution from members.
- Conserving 10 per cent development fund for maintenance of assets after completion of the project.

Following main features regarding community participation and management were observed in the best adjudged national watersheds:

(i) Grasslands were developed on a large scale by seeding of grass on common lands.
(ii) Cutting of dry grass was regulated from watershed area.
(iii) No one is allowed to graze cattle in the watershed area.
(iv) Community imposes and recovers penalty for violation of grazing regulations.
(v) *Ber* budding and plantation of trees was found very successful on hills and pastures.
(vi) Members of the society often meet to discuss issues relating to management of common property resources and overall development of the village.
(vii) Excess water from cultivated fields was diverted through channels.

Society is running smoothly with active people's participation to manage common property resources.

New approach of participatory natural resource management would enthuse the people, make the development programmes really useful for the local people, and make them come forward to welcome and participate in the great national venture towards sustainable management of the natural resources to produce abundance for our growing population without damaging the resource base.

Community Pumpsets

To provide irrigation facility to marginal and small ST farmers, this scheme was launched in tribal areas. Diesel pumpsets of 5 HP were provided free of cost. Maintenance of pumpset is the responsibility of group members. For this 50 per cent subsidy and 50 per cent bank loan are provided.

For sustainable development, it is essential that the benefit of schemes should continue uninterruptedly in the long run also. Hence, unless people's direct involvement is there at planning and execution stage, maintenance at future date after initial commencement of the project will continue to be a headache. Moreover, lakhs of amount spent to complete the project go waste if benefits cease to continue to poor beneficiaries. Hence, the people's participation and its importance in rural development programmes.

Terms and conditions were as under:

- Minimum land area should be 18 bighas in one block.
- Group should consist of 3-5 small or marginal farmers.
- They should own a well or a water source of river/nallah/ tank/anecut nearer the land to be irrigated.
- Diesel pumpset of 5 HP is provided free of cost along with 25 feet suction pipe and 25 feet delivery pipe of rubber. Total cost is around Rs 18,000.

If depth of nallah is more, remaining pipe is procured by member cultivators themselves.

A total of 93 persons intimately and directly connected with the lift irrigation programme were interviewed by research assistant and investigator personally. Results of such responses have been analyzed in depth.

- 71 per cent cultivators beneficiaries were small and marginal farmers. Average income before irrigation facilities was not more than Rs 2,000 per hectare.
- Maximum benefit was derived by big farmers whose irrigated area increased by 755 per cent as against small and marginal farmers whose area increase was 300 per cent and 250 per cent respectively, though all the farmers were tribals.
- Prior to scheme, no area was fully irrigated. It was either unirrigated or partly irrigated whereas after execution of lift irrigation schemes, 100 per cent area fully irrigated.
- Prior to scheme, *rabi* crop was of only gram and mustard which also was produced due to moisture retained in the soil. After the scheme, whole of the area is directly irrigated through lift water, well water or nallah/river waters.
- Farmers continued to grow cereal crops of wheat and barley only which were necessary either for their domestic consumption or sale in markets. No efforts were made to change the pattern to commercial crops like mustard, cotton, sugarcane, etc., which suited the soil and was also in demand.
- The analysis of survey shows that only 6 per cent cultivators gained the benefit of said crops. It also indicates indifferent attitude of farmers towards raising their income and standard of living through optimum utilization of irrigation facility.
- General cooperation of government officials was there. There was no beneficiary who complained about non-cooperation of officials. However, 87 per cent appreciated it but 13 per

cent felt that it was normal cooperation. All non-officials responded favourably and opined that cooperation of DRDA and Panchayat Samiti officials was very good in implementation of the programmes.

- Cultivators of course complained about indifferent attitude of agricultural supervisors and officers of agricultural department. Attitude of officials of state electricity board (RSEB) was also apathetic.
- As regards community participation, the analysis of this study revealed that cultivator beneficiaries became ready to contribute their 10 per cent share only after 10 to 15 extension meetings. Though community was enthusiastic in the beginning but it could not own the scheme. They continued to treat it as government scheme and find faults with it rather than running it successfully to their advantage.
- Some of the members defaulted in payment of their share of electricity dues or digging their canals or fixing outlet pipes. Default of few members resulted in whole community to suffer due to non cooperation in the scheme.

Lift Irrigation Schemes

As regards community participation in lift irrigation schemes, it was revealed that cultivator beneficiaries became ready to contribute their 10 per cent share only after 10 to 15 extension meetings. Though community was enthusiastic in the beginning but it could not own the scheme. They continued to treat it as government scheme and find faults with it rather than running it successfully to their advantage. Some of the members defaulted in payment of their share of electricity dues or digging their canals or fixing outlets pipes. Default of few members resulted in whole community to suffer due to non cooperation in the scheme.

Beneficiaries were well aware of the need for irrigation and they were eager to get the benefits but with minimum or no public contribution. Individual selfishness is more dominant than community feeling. Enthusiasm shown by the farmers at initial stage does not subsist long due to inter-member disputes due to shortage of power supply for irrigation of crops. They suggested that irresponsible members should not be included in the project to avoid disconnection for non-payment. Proper technical guidance should be continued as follow up by RSEB and DRDA officials for successful running of the schemes after initial commissioning. There were few ideal examples also where lift irrigation projects have been successfully running due to full cooperation of members.

Thus, it is clear from the analysis of the constraints identified in successful implementation of minor irrigation that active involvement and participation of the people is yet to take a concrete shape and form a movement. This calls for evolving a definite multi pronged strategy which may encompass awareness generation, resource mobilization and sustainable interest and enthusiasm as its key components. Needless to say that PRIs alone are capable of taking care of this strategy and ensuring its proper follow-up in the field. They can rely upon necessary support of official and voluntary agencies in their mission for rural development and upon people by imparting and cultivating a proper outlook among the beneficiaries who are still illiterate, poor and backward. It will be with a missionary mission alone that dream of Bapu's 'Gram Swarajya' can be realized.

Community Tubewells

Surface water has been managed by community through tank irrigation in the States of Karnataka and Andhra Pradesh since times immemorial but ground water management through community is of recent origin. In Uttar Pradesh, World Bank-aided Public Tubewells (PTWs) were designed with superior technology. Field water distributaries and feeders were dug. Electricity was assured for 16-18 hours. Agricultural extension service was also dovetailed with it. Farmers were involved in water distribution systems. PTWs are meant to benefit the marginal and poorer farmers. Tubewells are in clusters of 25 each in areas where economically backward people reside. The pattern of water distribution reflects the extent of farmers' participation in management of water resources from PTWs. Provision for farmers' participation was made at the layout stage itself. Benefits were quite gratifying.

In West Bengal also, many tubewells were turned over to panchayats. The panchayats, in turn, have turned them over to the user groups for actual management, though the accounting is under the supervision of the panchayats. This is said to be successful. Groups of farmers install pumpsets, and sell the water a little cheaper than the market rate. Though they are nominally supervised by panchayats but each group works independently.

In north Gujarat also, entrepreneur farmers get together and invest in tubewells, each of which irrigates 100-150 acres. Surplus water not required by members is sold out at Rs 18 an hour. There is a water market and sometimes a farmer has a choice of getting water from several outlets near his land. Another alternative model in Gujarat is that of tubewell companies. Five to 120 members having landholding in the command form an informal organization which is not registered. All promoters of the company enter into an agreement and elect a manager, open a bank account, all resources are internally generated. There is complete autonomy and self-governance, They run tubewells on community basis.

Managing Ground Water Together for Irrigation: A Case Study of Samraspur Village of Orissa

'Community Dugwell Irrigation Scheme' in a village of Cuttack district since 1971 is again a good example of people's participation. About 90 farmers having an area of 246.43 acres made a cooperative investment on 26 dugwells with the loan assistance of Rs 5,000 per family and subsidy from Land Development Bank and balance from their own manual labour. They got 33 per cent subsidy from DRDA and 25 per cent special state subsidy since 91 per cent beneficiaries were small and marginal farmers.

Resource base of water for irrigation of the lands of the cultivators was restructured. A total of 26 wells were fitted with electric pumpsets. Diseconomies associated with individual production units were scaled down because of cooperative venture in credit and marketing, etc. Besides successful maintenance and operation of irrigation water, group management system ensured effective as well as profitable use of water resource. The success factors were:

- Socio-economically homogenous class factor
- Urge for upliftment was a powerful stimulus
- Behavioural attitudes of cooperation

Cost was shared on area basis. Dry land became wet. Crop intensity increased. Multiple cropping could be possible. On-farm employment was provided. Productivity and income increased by 238.7 per cent (Rs 763.94 to Rs 1823.59 per acre). It was basically due to community participation and cooperation among beneficiaries.

Need for Change in Drinking Water Policy

The strategies of Rural Water Supply Programme hitherto adopted revolved around the basic premise that provision of safe drinking water is the responsibility of the government. All arrangements for construction and maintenance of drinking water works have so far been sole responsibility of the state. On the other side, water supply agency complains that the water rates were kept quite low, maintenance cost has increased due to high pay scales of public servants, cost of construction, pipes, motors and electricity is ever increasing and huge cash losses are incurred on account of water supply to people. Now, since population has increased tremendously and it is spread over a number of cities, towns and more than five lakh villages, it has become a herculian task for the state like Rajasthan to manage drinking water supply. Per capita supply of water has reduced even in cities and towns, and due to ever decreasing ground water levels, water is not supplied daily. There are towns which get water supply every second or even every fourth day. Huge funds are required to meet public demand for water specially in rural areas. Resources of the state are dwindling, working staff is also not sincere enough to attend to public complaints promptly and expectations of public are very high in a democratic system of governance.

Public water supply system is suffering from following problems:

- Charges for supply of water are too low. Hence, people use same water for cattle, for lawns, for flush latrines and miscellaneous washing purposes extravagantly.
- Government has no surplus funds for new investment in augmenting water supply sources as per increasing requirements of people. Whatever charges are recovered they are exhausted in salary bills of employees.
- Staff as well as public do not take any interest in preventing leakage of water from pipelines and taps.

- Government staff is not competent to manage the affairs efficiently.
- Rich people, influential elite, political bosses do not allow rates to increase to meet the deficit.
- Poor people do not get water because nobody cares for them. Leaders as well as bureaucrats are interested in huge commission from works rather than small villages.

Problems have arisen because of low water rates and maladministration of officials.

A change in the technology focus to handpumps fitted on tubewells and borewells had resulted in an impressive increase in total rural water supply coverage. However, the availability of potable drinking water in rural area, specially during the summer months, is still not satisfactory. Even though about one lakh habitations are covered every year, the number of problem habitations has not declined proportionately. Fast depletion of ground water level, heavy emphasis on new construction and poor attention to maintenance, non-involvement of people in operations and maintenance, neglect of traditional water management practices created necessity to involve community in drinking water management also so as to cover all rural habitations and reach the unreached with access to safe drinking water. Experience has shown that wherever women are involved in such consultation and in maintenance of the sources, the performance of supply systems has been effective.

Janta Jal Yojna

Rajasthan State initiated '*Janta Jal Yojna*' (People's Water Supply Scheme) in March 1994. It was experienced that governmentalization had been increasing day by day whereas no effort was made for participation of consumers in drinking water schemes. If participation of people increases in schemes, then success would definitely be more. 'Janta Jal Yojna' (JJY) could be sanctioned in villages where there is no source of drinking water or drinking water is insufficient as per norms of PHED.

- If any panchayat or village water committee proposes to make drinking water available to people from any well, PHED will reimburse electricity bill payable to the State Electricity Board in this behalf.

- Department can also deepen such well as per requirement.
- Department will also give grant for repairs of motor used for drawal of drinking water. In seven desert districts where water is deeper, motor is installed by the department itself.
- Sanctioning power was also decentralized. Sanction is to be accorded by district level committee consisting of collector and superintending engineer of PHED.
- There is no government operator for such schemes. Part-time operator is engaged by community. It can also be run on contract basis.
- Committee can also give domestic connections and it is authorized to recover monthly water rates from such consumers. Total income from 'JJY' belongs to village committee only.
- Such schemes can be sanctioned for villages having population of 500 and above.

As per progress report of PHED for the year 2002-03, a total of 1,894 JJY schemes are under operation by community, which include piped schemes (357), pump and tank schemes (879) and regional schemes (678).

Aapni Yojna

It is a global rural water supply project of unusual size and complexity. The project area covers about 20,000 kms. It is implemented by the community with the help of NGO partners under the guidance of German consultants. Community Participation Unit (CPU) was established in 1994 by a consortium of five leading NGOs with the aim to carry out Aapni Yojna (Our Scheme), an integrated rural water supply, sanitation and health education programme covering 956 villages and 11 urban areas of three districts of Churu, Jhunjhunu and Hanumangarh of Rajasthan State. It is funded by Government of India, Government of Germany through KFW and Government of Rajasthan. Present phase covers a population of 9 lakhs in 370 villages and two towns of Taranagar and Sardar Shahr.

Sustainable supply of water is ensured from Indira Gandhi Canal as groundwater is insufficient and saline. Project includes raw water storage from feeder canal, treatment plant, pumping stations and 1,100 MM pipelines, clear water reservoir, booster pumping station, elevated reservoir and cluster network of pipelines for different villages. Public stand posts and cattle water troughs are constructed in villages as part of the project.

Framework of participation is at various levels. First participation bond was between Government of Germany and Government of Rajasthan. Government of Rajasthan created special cell called Project Management Cell (PMC) from Public Health Engineering Department. Second participation bond was among leading five NGOs of Rajasthan for implementing community participation, sanitation, health education and women participation components of the project. Third participation bond is between the government and the NGO functionaries working closely with each other from down to top level, sharing each other's information and removing difficulties and working in a conducive environment. Fourth and important participation bond is between the project staff and the community. The regular interaction of CPU field staff with community has developed trust among the community for this project. The field intervention has empowered the village level community for managing their internal system independently. The communities are enthusiastically participating in all activities of the project.

Main Objectives

Community development through community participation and the implementation of activities as under:

- Reduction of wastage of water.
- Contribution by the community towards cost recovery.
- To create awareness and responsibility for the consumption of water (Water and Health Committee).
- To implement the construction of low-cost sanitation facilities.
- To launch women development activities and to promote family welfare measures.

Water Distribution Management Activities in a Village

Stepwise activities are as under:

- Environment building for project acceptance. Group meetings with males and females in every locality where role and responsibility of people is clarified.

- Village general meeting and WHC formation having proper representation of women and weaker sections. WHC consists of communicator, sanitation representative, payment collector, caretaker, and woman representative.
- WHC Orientation and job distribution clarifying their roles and responsibilities, maintaining records, etc.
- Social mapping for selecting sites for PSPs and CWTs.

Implementation Phase

- WHC conducts a census of the village, total family members and cattle for devising payment model. Census is updated every six months.
- Payment model for water charges according to which every household pays water charges as per the number of family members and cattle.
- Security money collection for one month advance is deposited with PMC.
- Site selection for PSP and CWT is done by WHC with the help of villagers. First women group selects the sites then WHC discusses and approves. PMC engineers check the technical feasibility and prepare layout for pipe laying and construction.
- Community arrange voluntary labour for trench digging. Project pays the money but it remains in corpus fund to be used for future maintenance.
- Construction work is done by PMC. Testing is done and after installation of village meter, the system is handed over to WHC.
- Community level training of WHC members is conducted regarding stand post attendants, water distribution management issues like record keeping, fault reporting, water charge collection and operation and maintenance of internal water supply system. A caretaker is trained for internal maintenance of the system and given one tool box from the project.
- In the follow-up phase, CPU monitors the proper use and maintenance of water supply installations.

- CPU monitors depositing of monthly bills regularly. In case any village fails to deposit, CPU analyses the problem and supports the villagers in solving the problem.

Project Benefits

- Availability of safe and potable water
- Availability for 24 hours
- Supply with adequate pressure
- Collection of water charges by Pani Panchayat
- O&M with Pani Panchayat
- No wastage of water
- 100 per cent recovery of water charges
- Women's empowerment
- Community participation

Aapni Yojna, the first of its kind in the desert area, employs an innovative approach to integrate communities in all phases of the project from planning, decision-making, construction, O&M phase to organization and maintenance of the whole scheme. Assurance of regular availability of water at the nearest point all the 24 hours, reduction in irresponsible use and wastage of water, the good image of CPU and active role of village committees and a well defined payment model are substantially contributing to the success of the project. It is recognized by the householders themselves who are prepared to invest large sums of money on maintaining "their own" facilities and regularly pay their water tariff. Even under severe drought, people's motivation to pay for water is not weakening at all. However, their ample understanding of the benefits makes all the difference. Beneficiaries are convinced of the necessity and advantages of safe drinking water and sanitation and its direct link to the reduction of water-borne diseases.

In Rajasthan 'Aapni Yojna' being implemented in district Churu is exemplary as regards community drinking water scheme. It is water democracy in action. It is a global rural water supply project of unusual size and complexity.

Sector Reforms and Swajaldhara Project

On the similar pattern are being implemented the Sector Reforms and Swajaldhara Project from April 1999 in rural water supply sector for institutionalizing community participation. It was mainly because water is today perceived by the rural public as a social right, to be provided free by the government, rather than as a scarce resource which must be managed locally as a socio-economic good in order to ensure its effective use. New policy is a transformation from a target-based, supply-driven approach which pays little attention to preferences of end-users, to a demand-based approach where users get the service they want and are willing to pay for is urgently required. Implementation of a participatory demand-driven approach will ensure that the public obtain the level of service they desire and can afford to pay. During review of Sector Reform Projects, the experience of community participation was not very much encouraging as only 29 per cent schemes could be completed and 24 per cent were handed over to community for running and maintenance.

Participatory and community-led 'Swajaldhara' is the latest experiment in community management of drinking water programme which was launched as a reform initiative throughout the country from 25 December 2002. The role of governments has to shift from service provider to that of facilitator for providing financial and policy support to the communities and their institutions for fulfilling the desired levels of services on a sustainable and equitable basis. There is increasing awareness of the need to strengthen technical and managerial capacity of different tiers of local governments.

Key Elements of Swajaldhara

1. Demand-driven and community participation approach.
2. Panchayats/communities to plan, implement, operate, maintain and manage all drinking water schemes.
3. Partial capital cost sharing by the communities upfront in cash and to be deposited with District Implementing Agency (DIA) before issue of 50 per cent as first instalment.
4. Full ownership of drinking water assets with Gram Panchayats.
5. Full operation and maintenance (O&M) by the users/ panchayats.

6. Beneficiary group should be a registered society and would submit proposal through Gram Panchayat or Block Panchayat as the case may be.
7. 10 per cent of the estimated capital cost of the schemes (5% in case of villages having 50% population of SC/ST as per 2001 census).
8. Contribution in cash is paid to the implementing agency (Zila Parishad).
9. DIA would be responsible for the formulation, implementation and management of the project.
10. DIA will select agencies and release the funds to Block Panchayat/Gram Panchayat/Village Water and Sanitation Committee (VWSC) and be responsible for proper management of funds.
11. DIA will sanction the scheme and enter into bipartite/tripartite agreements with executing agency.
12. One-time O&M grant will be given as revolving fund after running the scheme within six months. It will be on case to case basis.

Block Panchayat/Gram Panchayat/VWSC will be responsible for:

- Execution of sanctioned schemes;
- Place the progress of scheme implementation in each Gram Sabha meeting;
- Ensure community participation in scheme activities;
- Arrange community contribution towards capital cost in cash;
- Open and manage bank accounts for management of project funds;
- Procure construction material/goods and select contractors for construction activities;
- Supervise construction activities;
- After release of 50 per cent as first instalment, next instalment will be payable after 60 per cent fund of previous instalment has been utilized as per audited report;
- Commission and take over completed water supply works;
- Collect funds and manage O&M of water supply works; and
- Submit progress report to DIA within 10 days of the succeeding month.

Conclusion

It is essential that bureaucrats change their attitude of acting as facilitators for the community. Sincere efforts are required to build the capacity of user committees to shoulder the responsibility of management. NGOs will have to come forward to act as catalyst for empowering the stakeholders to manage community resources. Government will have to facilitate it through appropriate changes in rules and procedures and encourage NGOs for necessary interventions for the desired change. Leadership at the community level will also prove helpful in such paradigm shift.

▪ References

Centre for Science and Environment, New Delhi, Proceedings of the Conference on Potential of Water Harvesting, 3-5 October, 1998.

Conference of State Ministers of Rural Development on Rural Drinking Water, New Delhi, 5th December 2002.

Guidelines for Watershed Development (Revised 2001), Government of India, Ministry of Rural Development, Department of Land Resources, 2001.

Indian Reform Initiatives in Water Sector, Key Note Address by M. Venkaiah Naidu at Water Forum 2002, World Bank, Washington, 6th May 2002.

Joshi, L.K. and Rakesh Hooja, *Participatory Irrigation Management*, two volumes, Rawat Publications, Jaipur, 2000.

Joshi, R.P., *People's Participation in Lift Irrigation Projects in Tribal Areas of Rajasthan*, 1999.

Maloney, Clarence and Raju, K.V., *Managing Irrigation Together*, Sage Publications, New Delhi, 1994.

Mishra, G.P. and Bajpai, B.K., *Community Participation in Natural Resource Management*, Rawat Publications, Jaipur, 2001.

Participatory Planning of Watershed Programme, Part 11, Institute of Rural Management, Jaipur, 19

Rajiv Gandhi National Drinking Water Mission, Ministry of Rural Development, Guidelines for Implementation of Rural Water Supply Programme.

Ramakrishna Mission Lokashiksha Parishad, *West Bengal, Environment Development Through People's Participation.*

Ruparel River Flows Again, Tarun Bharat Sangh, Bhikampur, Alwar (Rajasthan).

Satish, S. and Sundar, A., *People's Participation and Irrigation Management*, Commonwealth Publishers, New Delhi, 1990.

Sector Reform Project Rajasthan – August 2002, Public Health Engineering Department Achievements 2002, Community Participation Unit, Aapni Yojana, PHED, Rajasthan.

Singh, Rajendra, *National Water Awareness Journey*, 2003.

Training Courses on Soil Conservation and Watershed Management (1994), Directorate of Soil Conservation and Watershed Management (NWDPRA), Department of' Agriculture, Government of Rajasthan, Jaipur.

Watershed Development in Rajasthan: Activities and Achievements, Watershed Development and Soil Conservation Department, Government of Rajasthan, 2002.

Chapter 7

Community Participation in Integrated Water Resource Management Policy Framework

M.V.S. Prasad

THE poor conservation outcomes that followed decades of intrusive resource management strategies and planned development have forced policy makers and scholars to reconsider the role of community in resource use and conservation. In a break from previous work on development which considered communities to hinder progressive social change, current writing champions the role of community in bringing about decentralization, meaningful participation, cultural autonomy, and conservation (Chambers and McBeth, 1992; Chitere, 1994; Etzioni, 1996). But despite its recent popularity, the concept of community rarely receives the attention or analysis it needs from those concerned with resource use and management.

We seek to redress this omission by investigating "community" in work concerning resource conservation and management. We begin by exploring the conceptual origins of the community, especially as it relates to writings on resource use. The ensuing analysis reveals that three aspects of community are most important to those who advocate a positive role for communities in water resource management – community as a small spatial unit, as a homogenous social structure, and as shared norms.

We suggest a more political approach. Community, we argue, must be examined in the context of conservation by focusing on the multiple interests and actors within communities, on how these actors influence decision making, and on the internal and external institutions that shape the decision-making process.

A focus on institutions rather than "community" is likely to be more fruitful for those interested in community-based natural resource management. We conclude by suggesting that research and policy move away from universalist claims either for or against community. Instead, community-based conservation initiatives must be founded on images of community that recognize their internal differences and processes, their relations with external actors, and the institutions that affect both.

Community in History

To understand the current widespread preoccupation with community requires an understanding of at least some history of the concept's use. Such a history shows the ways in which "community" has moved in and out of fashion, and prompts caution in accepting community as a panacea to problems concerning the conservation of natural resources.

Current perceptions of community appear strongly linked to analyses of 19th and early 20th century scholars attempting to understand the portentous transformations that rocked their world. The source of these changes was thought to lay in the economic sphere-industrialization, monetization, and production to satisfy material needs. Sir Henry Maine, for example, saw the world moving from relationships based on status, kin networks, and joint property to one based on contract, territory, and individual rights. Maine's underlying image of societal evolution influenced Tonnies's formulation of *Gemeinschaft* and *Gesellschaft*, or community and society. Tonnies's view of community as an organic whole continues to colour present conceptions to a significant degree, and accounts for some of the attraction community holds for many conservationists.

Most of these scholars of social change highlighted the disappearance of community and its replacement by other forms of social organization. Their theories of classification, in this sense, were also theories of evolution. For Marx and Engels, Spencer and Comic and even for Weber and Durkheim, society moved along an evolutionary path. Status, tradition, charisma, and religion would increasingly give way to equality, modernity, rationality, and a scientific temper. This theorization of social change automatically pits community against the market, since marketization and urbanization erode community.

Community and Conservation

Like more general works on community, the history of community in conservation is also a history of revisionism. Images of pristine ecosystems and innocent primitives yielded over time to views of despoiling communities out of balance with nature, mostly due to the double pronged intrusion of the state and market. A recuperative project on behalf of the indigenous and the local (community) has attempted to rescue community. But the rescue project has itself come under attack by new anthropological and historical research which suggests communities may not, after all, be as friendly to the environment. The practical and policy implications that accompany these changing images are immense.

The basic elements of earlier policy and scholarly writings about local communities and their residents are familiar. "People" were an obstacle to efficient and "rational" organization of resource use. A convincing logic under girded the belief that the goals of conservation and the interests of local communities were in opposition: Conservation required protection of threatened resources: wildlife, forests, pastures, fisheries, irrigation wood and drinking water. Members of local communities, however, rely on these resources for their fodder, fuelwood, water, and food and thus exploit them without restraint. This schematic representation, popularized by Garrett Hardin and bolstered by several theoretical metaphors that served to (mis)guide policy, provided a persuasive explanation of how resource degradation and depletion took place.

Empirical evidence about the context within which most rural communities are located helped prop up the view. The population of many rural areas in tropical countries has grown rapidly, even with out migration to cities. Demographic growth, it was argued, could only increase consumption pressures. Penetration by market forces, which linked local systems of resource use to a larger network of demand, further increased the pressure on natural resources. At the same time, many believed that poorly articulated and enforced property rights arrangements provided disincentives for individuals to protect resources.

While many of these beliefs persist, most of the current ideas about the community's role in conservation have changed radically: communities are now the locus of conservationist thinking. International agencies such as the World Bank, IDRC, SIDA, CIDA, Worldwide Fund for Nature, Conservation International, The Nature Conservancy, The Ford Foundation, The Mac Arthur Foundation, and USAID have all "found" community. They direct enormous sums of money and effort toward community-based conservation and resource management programmes and policies. A good of scholarly papers and policy-centric reports also feature community-based management (e.g., Arnold, 1990; Clugston and Rogers, 1995; Dei, 1992; Douglass, 1992; Perry and Dixon, 1986; Raju, Vaghela and Raju, 1993: Robinson, 1995). Exemplifying the swing toward community, a recent collection of essays on community-based conservation tells us, "Communities down the millennia have developed elaborate rituals and practices to limit off take levels, restrict access to critical resources and distribute harvests" (Western and Wright, 1994:1).

A host of other more specific factors have aided advocates of community-based conservation. The past several decades of planned development and top-down conservation practices have made one fact amply clear: the capacity of states to coerce their citizens into unpopular development and conservation programmes is limited. These limits are seen starkly when state actors attempt to discipline resource users. Where resources such as fodder, fuel wood, fish and wildlife are intrinsic to everyday livelihood and household budgets, even well funded coercive conservation generally fails. Faulty design, inefficient implementation and corrupt organizations have played an equally important role in the poor outcomes associated with state-centred policies.

Some contextual factors have also focused the attention of conservationists on community. With the spread of democratic political structures and the increasing insistence on participation, unrepresentative development and conservation projects have become as unattractive as they are impractical. The increasing prominence of indigenous and ethnic claims about the stewardship role of native populations in relation to nature (Clay, 1988; Redford and Mansour, 1996) assists those who advocate a central role for community. In addition, non-government organizations (NGOs) at different political levels have helped to amplify the voices of local, indigenous, and community groups (Borda, 1985; Borghese, 1987; Bratton, 1989a).

In light of the significant symbolic, theoretical, and intellectual resources available to advocates of community, it is somewhat surprising that claims on behalf of community based conservation often retain a rather simple quality. One such form such claims assume is that "communities" have a long-term need for the renewable resources near which they live, and they possess more knowledge about these resources than other potential actors. They are, therefore, the best managers of resources. Some refinements to this view can be found: if communities are not involved in the active management of their natural resources, they will use resources destructively (Sponsel, Headland and Bailey, 1996; Western and Wright, 1994). Still other work includes the notion of interests, in addition to that of needs: since it is in the interest of a community to protect its resources, it will. In its prescriptive form, this thesis of community-based conservation and resource management uses new beliefs about the suitability of communities to suggest policy recommendations. The implicit assumption behind these recommendations is that communities have incentives to use resources unsustainably when they are not involved in resource management. If communities are involved in conservation, the benefits they receive will create incentives for them to become good stewards of resources (if only the state and the market would get out of the way).

This vision of community – as the centerpiece of conservation and resource management – is attractive. It permits the easy contestation of dominant narratives that favour state control or privatization of resources and their management (Li, 1996). Such positive, generalized representations of community make available "points of leverage in ongoing processes of negotiation" (1996: 505, 509). But, such representations of community ignore the critical interests and processes within communities, and between communities and other social actors. Ultimately, such representations can undermine their advocates' long-term goal of increasing the role of community in natural resource management.

Role of Community in Water Resource Management

Much of the current discussion on community participation in water resource management, particularly with reference to countries in the Third World is limited. Clearly, an effort or movement to bring the entire populance into the sphere of community participation has to be made in order to create a more equitable and liveable environment, both in urban as well as rural areas. The idea that people should participate in planning, implementing and managing human settlements has gained acceptance among governments and development agencies. Arguments in favour of citizen's participation has been touted for long, and ultimately it means a readiness of both the government and the citizens to accept certain responsibilities and activities. It can also mean that the value of each group's contribution is seen, appreciated and used.

Preconditions to Participation

Past experiences on citizens' participation in various countries in Asia and Africa has clearly shown that participation cannot just happen; nor can it be taken for granted, either. There are several preconditions to participation which have to be l-net before it can be applied and sustained in a particular situation.

- Participation has to be a gradually developed response to an actual and pressing collective need of the citizens. This is need is used as a rallying point for the community to come together.

- The benefiting target group of a participative action has to be clearly defined, in order to utilize the common Interest in securing their position and improving their living conditions.
- It is of critical importance to inform the selected target groups, in a comprehensive manner, of all the relevant features of the programme. The aims, finance, technology, organization, management aspects have to be covered.
- In order that communication links between the authorities and the target group be maintained, there should be a strong community organization within the settlement, which could seek the assistance of an external NGO for information and motivation.
- A smooth functioning of the community organization structure ideally evolves through the collective efforts of the residents, with the aid of an accepted local leader. This is critical in representing the aspirations of the residents.
- The community leader and other members should be trained in the management process. Management is an important tool for reaching the desired aim of a self-help project, of keeping records and making responsible decisions in financial matters.

Essential Elements

The essential elements of participation requires that everyone understands, supports and is willing to join actively in the process. This requires much teaching and learning to be effective. There are many ways to increase participation in a water resource management, and some of those which are most useful to develop, monitoring and evaluation systems include:

1. *Use active rather than passive, and practical rather than theoretical methods:* Involve everyone, assign tasks which ensure everyone is involved or has a chance.
2. *Begin with an activity which is of interest to all:* Building a community map is a good start. Provide a simile outline. Each member can then put his or her house on the map. Other resources and landmarks can be added as required.
3. *Use small groups:* A large group intimidates the less bold, makes consensus more difficult, and inhibits spontaneity. Small homogeneous groups where there is mutual trust and

concern are more cooperative and supportive, at least initially.

4. *Provide meaningful data and information:* For those with no or little education, statistics and academic information cannot be interpreted. Simple graphic models, numbers and charts should be used.
5. *Facilitate access to more information:* Take the group to a library, a government office, school or on field trips where they can increase their knowledge base and learn where to go for information in the future.
6. *Conscientize the group:* Only when political awareness has been raised, are people willing and mentally able to help, themselves. Thus creation of awareness of the forces that oppress them is one of the most powerful forces for action, and collective action is necessary to achieve progress against oppression. Participation is the way to collective action.

In the end, the levels and effectiveness of people's participation depend largely on individual group organizers, leaders, teachers and facilitators. Not everyone is temperamentally suited to working with people in a participatory way, and not all can learn the skills which facilitate participation. Careful selection and effective training are both needed for people's participation to become a reality in development.

Essential Activities for Community Participation

The plea to establish a partnership between the state and the community comes with two crucial qualifications. Community based conservation can drastically change the shape of existing machineries. In the light of the above discussion of community participation, the following are the important activities of community based water resource management.

Identification of community needs

- Regular dialogue between government functionaries and communities before community needs are identified
- Using existing formal and informal organizations to identify community needs
- Training concerned government officials in communication skills and participatory approaches

Participation in project planning process

- Training for community leaders to plan community projects
- Changes in procedures to ensure bottom-up planning process
- Working with more than one group and organization during the planning process

Choice of location of services, beneficiaries and community leaders

- Involvement of community in choice of locations, beneficiaries, and leaders
- Working with more than one community
- Seeking community's consensus through community workers
- Training community leaders

Control over allocation of project resources

- Decentralization of financial and administrative authority
- Increased role of community development committee in allocation of project resources
- Training of community leaders in financial procedures and management

Mobilization of community resources

- Involvement of both formal and informal leaders in mobilizing community resources
- Providing more information to communities before asking them to contribute
- Seeking community contributions only about those activities affecting them directly and immediately
- Encouraging consensus in the community before the drive to mobilize community funds

Implementing project activities

- Standardization of procedures for project implementation under community control and supervision
- Provision of adequate training to community leaders and work concerning implementation of project activities

Periodic monitoring and evaluation of the project

- Giving the responsibility for the local project monitoring and evaluation to community development committee
- Training community leaders in methods of monitoring and evaluation of project performance

Actors and Interactions in Community Participation

To summarize, advocates of community based conservation forward a conceptualization of communities as territorially fixed, small, and homogeneous. These characteristics supposedly foster the interactions among members that promote desirable collective decisions. Figure 7.1 depicts the connections between different attributes of community and conservation outcomes indicated by the literature regarding community-based conservation. While certain types and levels of these characteristics might facilitate collective action, however, few studies demonstrate that this collective action is necessarily connected with conservation behaviour. Most important, few social scientists or policy makers have systematically tested these propositions in the field.

Figure 7.1
Conventional View of the Relationship between Community and Conservation

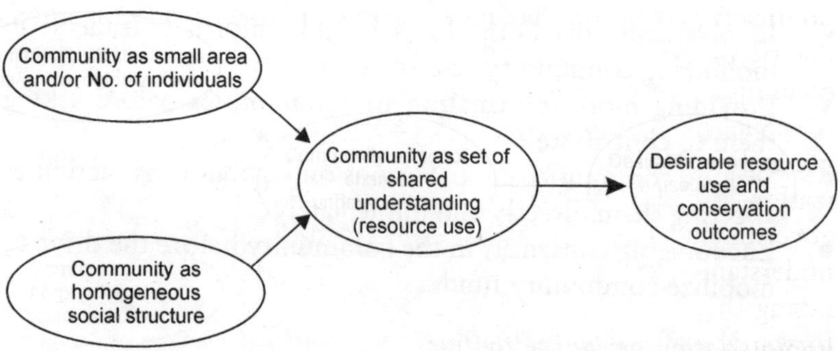

In fact, some community characteristics considered important to collective action may actually thwart conservation efforts. Small sized groups may be unable to defend their resources in the face of strong external threats, or be unable to manage resources if they are spread over large areas. Strongly held norms may support exploitative behaviour, or be resistant to outside attempts at their modification.

To be more accurate in our efforts to depict communities and their relationship with their natural resources – and thus to be more relevant to policy-making – we argue greater attention be focused on three critical aspects of communities: the multiple actors with multiple interests that make up communities, the processes through which these actors interrelate, and, especially, the institutional arrangements that structure their interactions. These three proposed foci for the study of community-based conservation allow for a better understanding of the factors critical to the success or failure of efforts aimed at local-level conservation:

(a) Multiple interests and actors

A growing number of studies that explore natural resource management at the local level do not find communities comprising just one group of individuals who possess similar endowments or goals. Instead, they find many subgroups; and within subgroups they find individuals with varying preferences for resource use and distribution. These authors bring to light the politics of the local: economic elites may vie with religious elites; chiefs may battle with their advisors; women may contest the rights of their husbands; the politically marginalized may dispute the acts of the politically dominant. Recognizing and working with the multiplicity of actors and interests is crucial for those advocating community-based programmes.

Such recognition indicates that empowering local actors to use and manage their natural resources is more than the decentralization of authority over natural resources from the central government to "a" community. The far more challenging task is to understand patterns of difference within communities. Recognizing that multiple actors exist at the local level is a useful step forward because it forces researchers to consider their different and dynamic interests. A more acute understanding of community in conservation can be founded only by understanding that actors within communities seek their own interests in conservation programmes, and that these interests may change as new opportunities emerge.

(b) Local-level processes

Within communities, individuals negotiate the use, management, and conservation of resources. They attempt to implement the agreed upon rules resulting from their negotiations. And they try to resolve disputes that arise in the processes of implementation of rules. These three types of local interactions are irreducibly influenced by the existing distribution of power and the structure of incentives within a given social group. Because the exercise of power and incentive-oriented behaviour are variable over time and space, and because all groups have members who can be strategic in their behaviour, planned conservation efforts can never address all contingencies completely.

Analyses of only local-level phenomena are insufficient to explain interactions at the local level. All local interactions take place within the context of larger social forces. Attempts by governments to implement community-based conservation and specific projects of NGOs that seek to involve communities are examples of directed influence on local level conservation. Such initiatives bring into the local context those larger political forces that generated the programmes. Other pressures changes in prices of deferent resources, development assistance, demographic shifts, technological innovations, institutional arrangements at deferent levels also impinge on local interactions.

Local interactions may also prompt responses from macro level actors. Local reactions to conservation programmes can lead to medications in the shape of these programs. Thus, although it is convenient to talk about the community and the state, or about the local and the external, they are linked together in ways that it might be difficult to identify the precise line where local conservation begins and the external (that helps construct the local) ends.

(c) Institutional arrangements

Institutions can be seen as sets of formal and informal rules and norms that shape interactions of humans with others and nature. They constrain some activities and facilitate others; without them, social interactions would be impossible (Bates, 1989; North, 1990). Institutions promote stability of expectations *ex ante,* and consistency in actions, *ex post*. They contrast with uncertain political interactions among unequally placed actors, and unpredictable processes where performances of social actors do not follow any necessary script. Strategic actors may attempt to bypass the constraints of existing institutions, and create new institutions that match their interests. But institutions remain the primary mechanisms available to mediate, soften, attenuate, structure, mold, accentuate, and facilitate particular outcomes and actions (Ensminger, 1992; Alston, Eggertsson and North, 1996; Agrawal, 1995b; Gibson, 1999). This holds whether change is radical, moderate, or incremental.

When actors do not share goals for conserving resources and are unequally powerful, as is likely the case in most empirical situations, institutions are significant for two reasons. On the one hand, they denote some of the power relations (Foucault, 1983: 222, 224) that define the Interactions among actors who created the institutions; on the other they also help to structure the interactions that take place around resources. Once formed, institutions exercise effects that are independent of the forces that constituted them. Institutions can change because of constant challenges to their form by the actions of individuals whose behaviour they are supposed to influence. No actual behaviour conforms precisely to a given institutional arrangement. Everyday performances of individuals around conservation goals possess the potential to reshape formal and informal institutions. Institutions can also change when explicitly renegotiated by actors. Institutions should be understood, therefore, as provisional agreements on how to accomplish tasks. Rather than setting the terms of interactions among parties with varying objectives, they help the behaviour of actors congeal along particular courses.

Authority to manage resources effectively at the local level requires the exercise of authority and control by local actors over three critical domains mentioned previously:

(i) making rules about the use, management and conservation of resources;
(ii) implementation of the rules that are created; and
(iii) resolution of disputes that arise during the interpretation and application of rules.

The authority to make rules defines who has the right to access, use, and conserve resources and exclude others from carrying out these activities. It also includes the determination of the ability to transfer these above rights. The authority to implement implies the rights and the abilities to meter and monitor the use of the resource, and specify sanctions against those who violate existing rules. The authority to resolve disputes includes the rights and capacities to ensure that sanctions are followed, and adjudicate in the case of disputes.

The problem of analyzing community-based conservation, thus, requires exploring a three step process of institutional formation. At each step, two issues must be addressed: Who will exercise the authority to make the rules? And, what will be the content of the rules? Typically, community-based conservation programmes devolve to local actors only while the authority to implement rules created elsewhere. Government agencies generally reserve for themselves the right to create rules and to arbitrate disputes.

In the light of the above discussion of multiple actors and interests and institutional arrangements around community participation, a different conceptualization of the relationship between different aspects of community and water resource management outcomes is possible. In contrast to Figure 7.1, the emphasis of this multiple interests are summarized in Figure 7.2. The figure does not present the theory of community-based water resource management, however, indicate some of the direction in which we can seek insight about the power of actors in community-based institutional water resource management.

Figure 7.2

An Alternative View of Community and Conservation of Water Resource

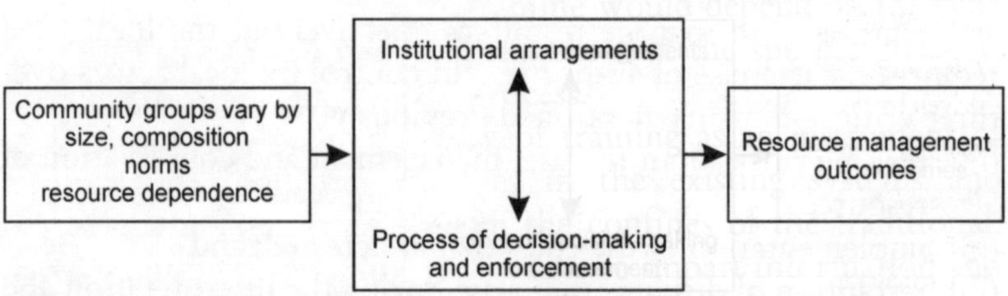

Checklist of Strategy Options for Incorporating Community

Table 7.1 broadly summarizes options for addressing community in designing and implementing water and sanitation sector interventions.

Table 7.1
Options for Addressing Community

Levels: Objective/Options	*Key Stakeholders*
Level: Country Policy Objective: Develop and implement more efficient cost-effective, and demand-responsive water and sanitation policies by incorporating gender issues. *Options:* • Introduce gender issues in sector reviews, policy workshops, and other activities that are part of policy development. • Put gender issues on the agenda of annual sector meetings and policy implementation reviews • Include gender expertise on policy development and implementation teams.	Government ministries, donor agencies, women's and other NGOs, and sometimes user groups
Level: National Water Programmes *Objective:* Improve country-level program design and implementation by incorporating gender concerns. *Options* • Include community in country programme framework. • Include gender-related guidelines and principles in country programme. • Employ community analysis in designing projects. • Include government staff with community expertise in monitoring the national programme. • Monitor community issues regularly.	Government ministries, donor agencies, women's and other NGOs, and user groups
Level: Water and Sanitation Projects *Objective:* Design and implement projects that are driven by the demands of community. *Options* *Project Design* • Structure project rules and procedures to facilitate participation by both men and women. • Determine community roles in the sector in the proposed project area. • Determine barriers to community-appropriate project implementation.	Project staff, local government, and user groups

Contd...

Contd...

Levels: Objective/Options	*Key Stakeholders*
• Determine steps to reducing or removing the barriers. • Make projects flexible so they may adapt appropriately as more is learned about gender issues. • Include a community expert on the team during project design/preparation. *Implementation and Supervision:* • Amend project rules and procedures as needed to facilitate participation by both men and women in implementation. • Ensure that project management is aware of the importance of gender issues through training, workshops, and study tours. • Include community experts on project. implementation staff. *Monitoring and Evaluation (M&E)* • Collect, tabulate, and analyze indicators by community as appropriate. • Include specific indicators addressing community issues in project M&E systems. • Examine community-related M&E indicators during supervision.	
Level: Community *Objective:* Increase project sustainability by improving implementation at community level. *Options:* *Project Design* • Base men's and women's involvement on the local cultural context: for example, separate meetings of men and women or female staff meeting with community women, where necessary. • Use participatory techniques to ensure both women's and men's participation in project decision-making concerning: • Technology choice • Cost recovery • O&M arrangements	Project staff, community members, and women's and other NGOs

Contd...

Contd...

Levels: Objective/Options	*Key Stakeholders*
• Obtain men's and women's preferences about	
• Technology design	
• Siting of facilities.	
Operations and Maintenance	
• Suggest that a certain percentage of water and sanitation committee members be women.	
• Suggest that women should hold at least one water and sanitation officer post, such as treasurer	
• Provide training for both men and women in the roles they are to fill in the project.	
• Include additional training for women in leadership and organization, as appropriate.	
• Train both women and men in basic O&M techniques.	

Conclusion

To analyze the community-based water resource management, this paper began by casting a critical historical eye at the notion of community. Current works on community borrow extensively, if unconsciously, from past writings. Visions of community as an organic whole, as small and territorially fixed, as under siege and eroding, or as standing in opposition to markets and states, can be traced directly to writings from the 19th and the early 20th century. A longer term perspective on community prompts caution before one embraces it as a general answer to conservation-related woes.

The celebration of community is a move in the right direction. But the implications of turning to it are little analyzed in most writings on community-based conservation. The existing literature on community-based conservation reveals a widespread preoccupation with what might be called "the mythic community": small, integrated groups using locally evolved norms to manage resources sustainably and equitably.

Such characteristics capture the realities of few, if any, existing communities. The vision of "the mythic community" fails to attend to differences within communities. It ignores how differences affect processes around conservation, the deferential access of actors within communities to various channels of influence, and the possibility of "layered alliances" spanning multiple levels of politics. Small, territorially attached, and relatively homogeneous communities, where they exist, might find it easy to make decisions collectively.

Our advocacy is for a changed emphasis for those of us who believe in locally-oriented management of resources and a move away from states and markets. Greater autonomy to local groups means that external actors would have to relinquish control over the rules and the outcomes of community-based conservation. In addition, the directions in which institutional outcomes in local spaces will unfold cannot be plotted precisely, they can only be roughly assessed. Demands for greater certainty suffer from the same utopian longings that identify "community as shared norms" as the solution to problems of conservation.

We conclude our analysis by discussing four possible areas for new research. In the preceding text of the essay we have only hinted at each of these following four points. They require considerable more development. We state them here as issues for future work.

First, community-based conservation would more profitably be founded on principles of checks and balances among various parties local groups, government actors, even NGOs and aid agencies rather than faith in the regenerative capacities of any one of them. Unchecked authority for community-level decisions is likely to lead to perverse conservation outcomes.

Second, local groups are usually the least powerful among the deferent parties interested in conservation. Community-based water management requires, therefore, that its advocates make more strenuous efforts to channel greater authority and power toward local groups. Only then can such groups form effective checks against arbitrary actions by governments and other actors. Critical to such attempts is the need to forge federated structures of community user groups that can negotiate with government officials and aid agencies on more equal terms than those prevailing today. Negotiations on terms of equality are foundational to hold government actors accountable.

Third, those interested in community-based conservation should seek to implement reasonable processes of decision-making rather than focus upon guarantees about outcomes. "Reasonable" implies that (a) different interests, especially those that are usually marginal, are represented in decision-making, (b) mechanisms exist to ensure that the outcomes of current decision processes are going to form part of the data on which future decisions will be based, and (c) the performance of those who make decisions is periodically reviewed by those affected by decisions. Local representatives of communities, and those elected as officials in federated structures of community groups must themselves be accountable to their constituents if a new understanding of community-based conservation is to have any teeth.

Regular and open elections in which decision makers submit to choices made by their constituents may be indispensable to ensure such accountability. Without mechanisms of accountability, federations of community groups may become yet another channel for centralizing tendencies.

Finally, effective institutionalization of community-based conservation requires that local groups have access to adequate funds for implementing the rules they create. The sources for these funds should also be local, raised through contributions of users rather than granted by central governments. Over time this would mean that government agencies not just cede their authority to make rules about conservation, but that community groups also demand control over the resources themselves.

The points outlined above do not provide a blueprint for community-based conservation. Rather, they emphasize the importance of institutions, the ubiquity of political processes, the need to institute cheeks to contain arbitrary exercise of power, and the impossibility of escape from an uncertain future.

■ References

Agrawal, A. and Yadama, G. (1997) How do Local Institutions Mediate Market and Population Pressures on Resources? Forest Panchayats in Kumaon, India, *Development and Change* 28 (3), 435-65.

Arnold, J.E.M. (1990) Social Forestry and Communal Management in India. *Social Forestry Network Paper* 11b, Overseas Development Institute, London.

Ascher, W. (1995) *Communities and Sustainable Forestry in Developing Countries.* ICS Press, San Francisco.

Bender, T. (1978) *Community and Social Change in America.* Johns Hopkins University Press, Baltimore.

Chambers, R.E. and McBeth, M.K. (1992) Community Encouragement: Returning to the Basis for Community Development. *Journal of the Community Development Society* 23 (2), 20-38.

Chitere, O.P. (ed.) (1994) *Community Development: Its Conceptions and Practice with Emphasis on Africa.* Gideon S. Were Press, Nairobi.

Coleman, J.S. (1990) *Foundations of Social Theory.* Harvard University Press. Cambridge, MA.

Dove, M. (1982) The Myth of the "Communal" Longhouse in Rural Development. In: *Too Rapid Rural Development,* ed. by C. MacAndrews and L.S. Chin, pp. 1478. Ohio State University Press, Athens.

Douglass, M. (1992) The Political Economy of Urban Poverty and Environmental Management in Asia: Access, Empowerment and Community Based Alternatives. *Environment and Urbanization* 4 (2), 9-32.

Etzioni, A. (1996) Positive Aspects of Community and the Dangers of Fragmentation. *Development and Change* 27, 301-314.

Fernandes, A. (1987) NGOs in South Asia: People's Participation and Partnership. *World Development* 15 (Supplement) 39-49.

Ghai, D. (1993) Conservation, Livelihood and Democracy: Social Dynamics of Environmental Change in Africa. *Osterreichische Zeitschrift fur Soziologie 18,* 56-75.

Hoban, T.J. and Cook, M.G. (1988) *Challenge of Conservation*. Forum for Applied Research and Public Policy 3. 100-102.

Li, T.M. (1996) Images of Community: Discourse and Strategy in Property Relations. *Development and Change* 27 (3), 501-28.

Murphree, M.W. (1993) *Communities as Resource Management Institutions* International Institute for Environment and Development, London.

Myers, N. (1991) The World's Forests and Human Populations: The Environmental Interconnections. In: *Resources, Environment, and Population: Present Knowledge, Future Options,* ed. by W. Davis and M. Bernstam, pp. 237-251. Oxford University Press, New York.

North, D. (1990) *Institutions, Institutional Change, and Economic Performance.* Cambridge University Press, Cambridge.

Ostrom, E. (1990) *Governing the Commons: The Evolution of Institutions for Collective Action.* Cambridge University Press, Cambridge.

Ostrom, E. (1992) *Crafting Institutions for Self-Governing Irrigation Systems.* Institute for Contemporary Studies Press, San Francisco.

Ostrom, E.E., Schroeder, L. and Wynne, S. *(1993) Institutional Incentives and Sustainable Development: Infrastructure Policies in Perspective.* Westview Press, Boulder, CO.

Parsons, T. (1951) *The Social System.* The Free Press, New York.

Perry, J.A. and Dixon, R.K. (19861) An Interdisciplinary Approach to Community Resource Management: Preliminary Field Test in Thailand. *Journal of Developing Areas* 21 (1), 31-47.

Peters, P. (1994) *Dividing the Commons: Politics, Policy and Culture in Botswana.* University of Virginia Press, Charlottesville.

Rajasekaran, B. and Warren, D.M. *(1994)* Socioeconomic Development and Biodiversity Conservation: The Kolli Hills. *Indigenous Knowledge and Development Monitor* 2 (2).

Raju, G., Vaghela, R. and Raju, M.S. *(1993) Development of People's Institutions for Management of Forests.* Ahmedabad, India: Viksat, Nehru Foundation for Development.

Western, D. (1994) Ecosystem Conservation and Rural Development. In: *Natural Connections: Perspectives in Community-based Conservation,* ed. by David Western and R. Michael Wright, pp. *15-52.* Island Press, Washington, DC.

Western, D. and Wright, Michael R. (eds.) *(1994) Natural Connections: Perspectives in Community-based Conservation.* Island Press, Washington, DC.

Wilmsen, E. (1989) *Land Filled with Flies: A Political Economy of the Kalahari.* University of Chicago Press, Chicago.

Chapter 8

Water Scenario in Rajasthan

Policy Potential and Problems

A.S. Kapoor

THE total amount of water on the earth is about 1,400 million cubic kilometres, of which 97 per cent is in oceans, which cover about three-fourths of the earth surface. Fresh water available for human use is less than 1 per cent of the total water on the earth. The global water balance is shown in Figure 8.1.

The average annual precipitation over India is 400 million hectare metres (m.ham). It is envisaged that 70 m. ham of surface water and 35 m. ham of groundwater, total 105 m. ham can be mobilized for human use by the year 2025. Rajasthan is the largest state in the country covering 10.4 per cent of the country's area. The state's population is 5.65 per cent, the surface water 1.11 per cent and the groundwater 1.72 per cent of the corresponding features of the country.

The per capita availability of water in Rajasthan is one-fourth of that of the country's average and availability per unit of geographical

Figure 8.1

Global Water Balance and Fluxes (Todd, 1970)

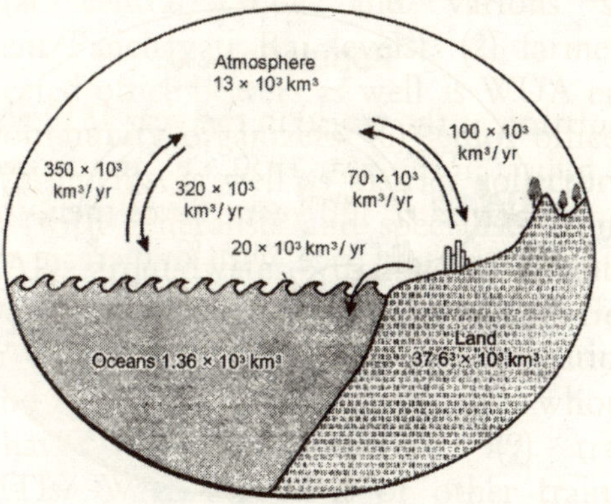

Figure 8.2

Water Resources in Rajasthan

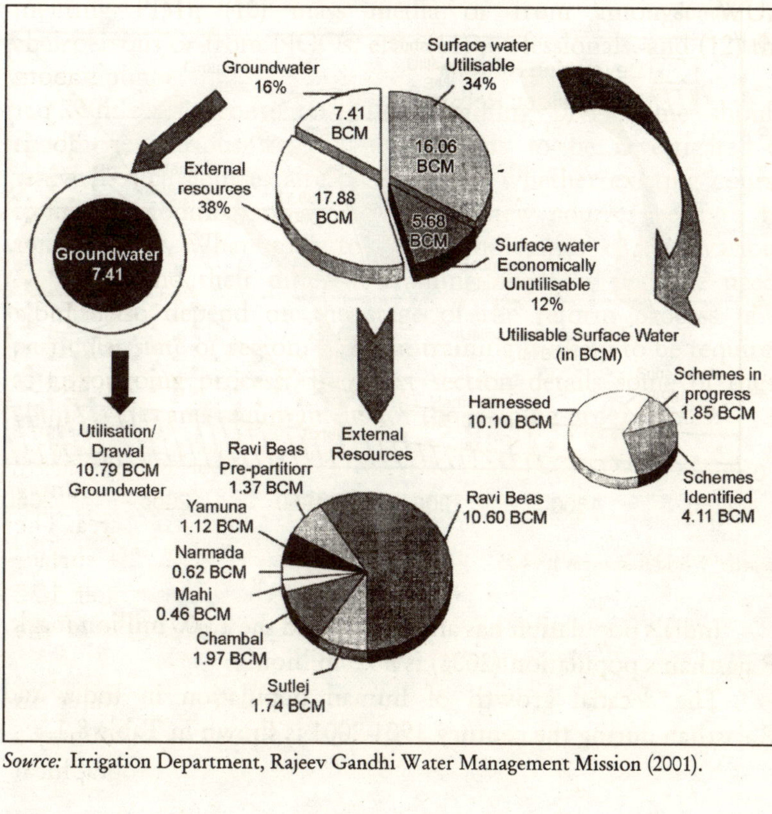

Source: Irrigation Department, Rajeev Gandhi Water Management Mission (2001).

area is only about one-tenth. The overall position of water resource in Rajasthan is shown in Figure 8.2.

Population

The total population of the world in the year 1810 was 1 billion. It increased to 2 billion in the year 1930, i.e., an increase of 1 billion occurred during a period of 120 years. Since then, the population has been increasing at a rapid pace, with an increase of last 1 billion (from 5 to 6 billion) occurring in a period of just 11 years. The past trend and future projection in world population growth are shown in Figure 8.3.

Figure 8.3

Trends & Projections in World Population Growth: 1750-2050

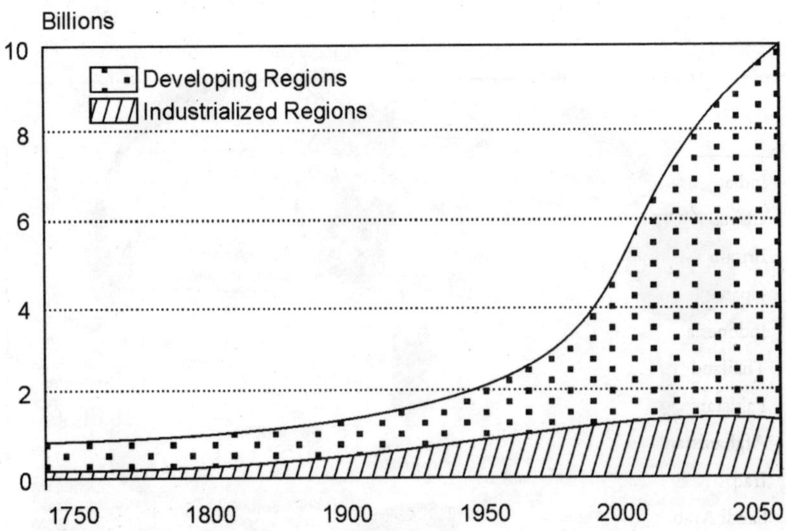

Source: World Resources 1994-95

India's population has already crossed the 1,000 million mark. Rajasthan's population (2001) is 56.5 million.

The decadal growth of human population in India and Rajasthan during the century 1901-2001 is shown in Table 8.1.

Table 8.1

Population in India and Rajasthan

(in million)

Year	*1901*	*1911*	*1921*	*1931*	*1941*	*1951*	*1961*	*1971*	*1981*	*1991*	*2001*
Rajasthan	10.3	11.0	10.4	11.8	13.0	15.9	20.1	25.8	34.2	44.0	56.5
India	238.4	252.1	251.3	279.8	318.7	361.1	439.2	548.2	683.3	846.3	1027.0

Source: Rajasthan Patrika, 22 June 2003.

According to the United Nations estimates of future population projections, in another 30 years, India would be the most populated country in the world. Population growth projections of some countries are shown in Table 8.2:

Table 8.2

Population Growth Projections of Some Countries

(in million)

Country	*Year*		
	2000	*2025*	*2050*
India	1002.0	1363.0	1628.0
China	1264.0	1431.0	1369.0
Britain	59.7	64.0	64.1
Japan	120.6	120.9	130.4
Indonesia	210.2	270.3	310.1
Thailand	62.0	72.1	73.3
Pakistan	150.6	227.0	285.0
Afghanistan	26.6	48.0	76.2
Iraq	23.1	41.0	54.9
Saudi Arabia	21.6	39.9	54.4
Yemen	17.0	38.6	69.2
Australia	19.2	22.8	24.9
Brazil	171.0	221.0	244.0
USA	275.0	337.0	403.0

Source: Rajasthan Patrika, 22 June 2003.

Water Requirement and Availability

Rajasthan's annual utilization of water (1995) was 17.45 BCM. The projected figures of water utilization for the years 2015 and 2045 are 23.34 BCM and 27.71 BCM respectively. The projected population of the state for the year 2015 is 72.9 million and for the year 2045 it is 102.9 million. The position of estimated annual water availability per capita emerges as shown in Table 8.3:

Table 8.3

Per Capita Annual Availability of Water – Rajasthan

Year	*Population (million)*	*Annual Availability of Water (estimated)*		*Annual Requirement of Water @ 500 m3 (BCM per capita BCM*
		Total (BCM)	*Per capita (BCM)*	
1995	48.6	17.65	363	24.30
2015	72.9	23.34	320	36.45
2045	102.9	27.71	269	51.45

Source: Irrigation Department, Government of Rajasthan, Vision 2045 (September, 2000).

Water scarce conditions prevail when and where the annual water availability per capita falls below 500 cubic metres. In Rajasthan, water scarcity is quite acute and is likely to get progressively worse with passage of time. The primary cause of the water shortage problem is population growth. If, somehow, the population of the state could be frozen at the current level of 56.5 million the per capita annual availability of water in year 2045 can be 500 cubic metres. The position is illustrated in Figure 8.4.

It may not be feasible to stop population growth altogether. But, in absence of commitment and without effective measures to control population growth, water scarcity problem in the state cannot be resolved.

If countries with such diverse history, culture and polity as China and Great Britain can check and control their population growth, why can not India do so? Scandinavian countries are reported to have already achieved sub-zero population growth rates. The reasons of poverty, illiteracy and compulsions of a democratic system, which are often advanced to explain the tardy

Figure 8.4
Per Capita (Annual) Availability of Water – Rajasthan

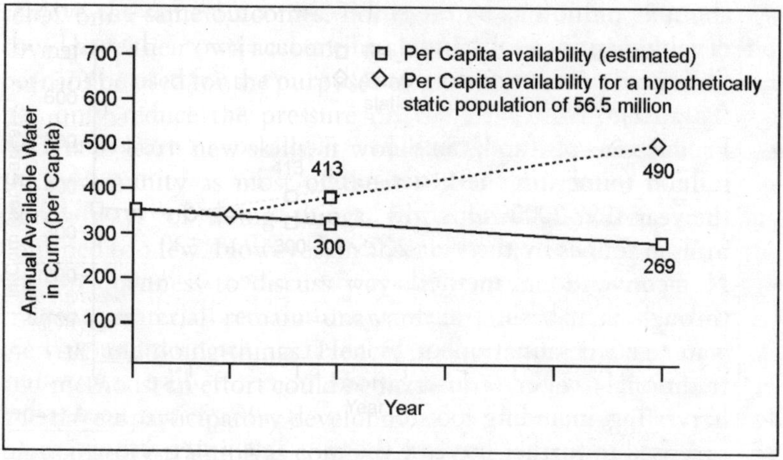

progress in population control measures should not be acceptable. What is in larger public interest, should be attempted vigorously and must be achieved. The foremost necessary step to overcome the problem of water scarcity in India and particularly in Rajasthan, therefore, is to control population growth without which regardless of what we may do, not more than half success would be achievable.

The National Water Policy

The National Water Resources Council, for the first time, adopted the National Water Policy in the year 1987, which was modified and titled as "National Water Policy, 2002". Some relevant extracts from the Policy Paper are reproduced here:

- Water is part of a larger ecological system. Realising the scarcity attached to the fresh water, it has to be treated as an essential environment for sustaining all life forms (Para 1.3).
- ... As the country has entered the 21st century, efforts to develop, conserve, utilize and manage this important resource (water) in a sustainable manner, have to be guided by the national perspective (Para 1.4).
- ... Complex issues of equity and social justice in regard to water distribution are required to be addressed. ... (Para 1.6).

- ... While the gross irrigation potential is estimated to have increased from 19.5 million ha at the time of independence to about 95 million ha by the end of the year 1999-2000, further development of a substantial order is necessary if the food and fibre needs of outgrowing population are to be met with .... (Para 1.7).
- Production of foodgrains has increased from around 50 million tonnes, in the fifties to about 208 million tonnes in the year 1999-2000. This will have to be raised to around 350 million tonnes by the year 2025 AD ... (Para 1.8).
- Non-conventional methods for utilization of water such as through inter-basin transfers, artificial recharge of ground-water and desalination of brackish or sea water as well as traditional water conservation practices like rainwater harvesting, including roof-top rainwater harvesting need to be practised to further increase the utilisable water resources. .... (Para 3.2).
- Water should be made available to water short areas by transfer from other areas including transfer from one river basin to another, based on a national perspective, after taking into account the requirements of the areas/basins (Para 3.5).
- In the planning and operation of systems, water allocation priorities should be broadly as follows:
  - Drinking water
  - Irrigation
  - Hydropower
  - Ecology
  - Agro-industries and non-agricultural industries
  - Navigation and other uses
- The involvement and participation of beneficiaries and other stake holders should be encouraged right from the project planning stage itself (Para 6.8).
- Besides creating additional water resources facilities for various uses, adequate emphasis needs to be given to the physical and financial sustainability of existing facilities. There is, therefore, a need to ensure that the water charges for various uses should be fixed in such a way that they cover at least the operation and maintenance charges of providing the service initially and a part of the capital costs subsequently

These rates should be linked directly to the quality of service provided. The subsidy on water rates to the disadvantaged and poorer sections of the society should be well targeted and transparent (Para 11).

- ... Water Users Associations and local bodies such as municipalities and 'Gram Panchayats' should particularly be involved in the operation, maintenance and management of water infrastructures/facilities at the appropriate levels progressively, with a view to eventually transfer the management of such facilities to the user groups/local bodies (Para 12).
- Private sector participation should be encouraged in planning, development and management of water resources projects for diverse uses wherever possible. Private sector participation may help in introducing innovative ideas, generating financial resources and introducing corporate management and improving service efficiency and accountability to users. Depending upon the specific situations, various combinations of private section participation, in building, owning, operating, leasing and transferring of water resources facilities, may be considered (Para 13).
- Efficiency of utilization in all the diverse uses of water should be optimized and an awareness of water as a scarce resource should be fostered. Conservation consciousness should be promoted through education, regulation, incentives and disincentives (Para 16.1).
- The water sharing/distribution amongst the states should be guided by a national perspective with due regard to water resources availability and needs within the river basin. Necessary guidelines including for water short states even outside the basin, need to be evolved for facilitating future agreements amongst the basin states (Para 21.1).
- The Inter-State Water Disputes Act of 1956 may be suitably reviewed and amended for timely adjudication of water disputes referred to the tribunal (Para 21.2).
- There is an urgent need of paradigm shift in the emphasis in the management of water resources sector. From the present emphasis on the creation and expansion of water resources infrastructures for diverse uses, there is now a need to give

greater emphasis on the improvement of the performance of the existing water resources facilities. Therefore allocation of funds under the water resources sector should be re-prioritized to ensure that the needs for development as well as operation and maintenance of the facilities are met (Para 22).

- Formation of Water Users Association with authority and responsibility should be encouraged to facilitate the management including maintenance of irrigation system in a time bound manner (Para 23.3).

The State Water Policy

The State Water Policy document of the Government of Rajasthan is on similar lines as the National Water Policy. Of particular importance, it incorporates the following:

- The overwhelming interest of the state is to bring, by physical and managerial measures, as much of the potentially available resources into beneficial utilization as is physically and economically feasible (sub-para 3, Maximizing Water Availability).
- The case of full utilization of states share in Ganga waters shall be pursued (sub-head 3, Maximizing Water Availability, item (k).
- Water rates shall be increased and collections shall be improved (sub-head 5, Maintenance and Modernization, item (b), also sub-head II Water Rates)
- Water Users Associations shall be encouraged to undertake maintenance, repairs and modernization of works (sub-head 5, Maintenance and Modernization, item (e), also sub-head 12, Participation of Water Users).
- Farmers shall be encouraged to adopt high efficiency water equipment and practices and use of groundwater in conjunction with surface water (sub-head 10, Irrigation Water, item (b).

The National and State Water Policies cover almost all aspects on water. The water resource potential and constraints are well known. Achievements and shortcomings are identified. What should be the future strategy and programme are well stated. What the policy papers do not describe are the causes and reasons that have led to the shortcomings, and how the impeding obstacles would be overcome. Another weakness of the policy papers is that they do not identify the factors that are most responsible for less than satisfactory utilization of the water resource. There is no prioritization. There is no identification and categorization of most important activities and programmes, that should be the first focus of attention,

The problem of population growth that is subjecting the limited water source to increasing pressure has been discussed earlier. Before dwelling on other problems and possibilities, let us have a look at the past achievements and present status.

Present Status

In the year 1947, there were in all 44 major and medium projects and 2,272 minor irrigation works in the state with an irrigation potential of 4 lakh ha. In the year 2000, there were 104 major and medium projects and 4,786 minor irrigation works in the state, with an irrigation potential of 28.12 lakh ha. The position of this development is illustrated in Figure 5.

Figure 5

Irrigation Development in Rajasthan

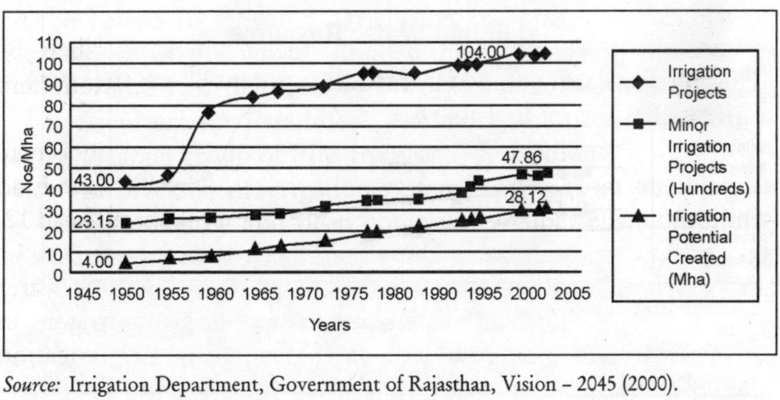

Source: Irrigation Department, Government of Rajasthan, Vision – 2045 (2000).

For groundwater use, the state has been, divided into 237 blocks. There is overexploitation of groundwater as a result of which the groundwater table is falling and the quality of the groundwater is deteriorating. Out of the total rural habitations of

93,946 (villages – 37,889, 'dhanis' – 56,057), 18,690 are fluoride affected and 16,344 have salinity problem. In 26 districts out of a total of 32, the groundwater level during the years 1998 to 2001 went down between 2 to 8.66 m. In comparison, the groundwater level had gone down by only 1 m in the preceding 14 years. The comparative position is as shown in Table 8.4.

Table 8.4

Groundwater Blocks: Status of Exploitation

Number of Blocks	*Year*	
	1998	*2001*
Total number of blocks in the state	237	237
Overexploited blocks (most critical)	41	86
Dark blocks (critical)	26	80
Gray blocks (semi-critical)	34	21
Safe blocks	135	29

There is some scope of developing the surface water potential in the state but hardly any of developing the groundwater potential, which is already overexploited.

Limited Water Resource

The ultimate annual water resource potential of the state is estimated as shown in Table 8.5:

Table 8.5

Water Potential in Rajasthan

Source	*Potential (BCM)*
Surface water (utilisable)	16.05
Groundwater	7.41
External sources	17.88
Total	41.34

In addition, there is a surface water resource of 5.66 BCM which is economically unutilisable. Out of the 16.05 BCM utilisable water potential, 10.10 BCM stands harnessed, works are in progress for 1.85 BCM (year 2000) and schemes have been identified for 4.11 BCM. Groundwater potential is already overexploited. There is not much that can be done to significantly increase the water resource potential within the state.

National Project of Interlinking the Rivers

A major contribution to the State Water Resource can come from the interlinking river project, that has come in focus recently. The inter-state dispute between Karnataka and Tamilnadu, and the Supreme Court directive enlivened interest in the project. The Prime Minister has constituted a 'Task Force' for this gigantic project. How can this project help Rajasthan? Surplus water in the country is mostly in the eastern region. By construction of storage dams and links between rivers the surplus water can be transferred from east to west as well as, from north to south. How much water is surplus and how much can be economically transferred are subjects that would require careful study. A preliminary study made a few years back by the National Water Development Agency (NWDA) contemplated a transfer of about 3 BCM of water to Rajasthan/Gujarat through Sharda-Yamuna links. Kapoor (1971), in a paper, suggested that this quantity should be 58 BCM for Rajasthan. It is easy to see the enormity of stake for the state. The handicap of water scarcity in Rajasthan can be wholly overcome for all times, through inter-basin transfer of sufficient quantity of water towards the west.

We in Rajasthan have yet to fully comprehend the magnitude and impact of the proposition? No other project or programme in any field can bring comparable benefits to the state. It may take a very long time for the project to materialize but it is of utmost importance that the state does all and everything that is needed to secure maximum possible share in the water that can possibly become available from the project. Proactive efforts are needed at all levels. These would include study and formulation of feasible alternative propositions; economic, social and environmental evaluations, assessment of technical and physical problems likely to be encountered. Project reports, discussion papers and material for creating public awareness should be prepared. Objections, concerns and opposition expressed against the scheme, which are already being widely voiced in the media, would need to be carefully considered, accounted for and countered.

The government and the people of the state should treat this project as the most important item on the development agenda and start a crusade in right earnest. To begin with all available literature on the project should be obtained and studied. Links and liaison should be established with the 'Task Force', the NWDA, and the concerned Central Government ministries and agencies. A programme of preparing articles, media-stories, TV films should be taken in hand. Serious papers on social, economic and ecological aspects should be written by experts and publicized. Technically feasible proposals would need to be prepared and projected by competent persons.

The Low Water Use Efficiency

Presently, about 93 per cent of water is used for agriculture. Because of the increasing demand of water for domestic and other uses, the percentage of water use for agriculture is expected to come down to 80 per cent by the year 2045. The water use efficiency of surface water, is quite poor. The Irrigation Department estimates the overall efficiency as less than 30 per cent. Water is lost by evaporation in storage reservoirs, through seepage from the canals and by infiltration from the fields during irrigation. Some of these losses are inevitable but with proper management and adopting water saving measures like lining of canals and with sprinkler/drip system, the efficiency of water use can be improved to 70 per cent. The total volume of water that can become available by savings in this manner can be about 10 BCM. This is possible but would require investments in upgrading the irrigation infrastructure and reforms in the administrative and management systems.

Inadequate Financial Resources

Lack of financial resource is a constraint faced by the state all the time. The demand on available resources keeps on growing. The competing demands are for development sectors like agriculture, power industry, highways, irrigation, etc. Social services sector like education, health, etc., need large allocations. Famine and droughts, which occur quite frequently, upset the development programmes. The problem can be best managed by (a) generating additional revenue, and (b) judicious distribution of available resource, priority being given to the sectors that give best economic returns.

Attempts to generate additional revenue by new taxes or revising the rates of service charges or reducing the subsidies always meet opposition and all governments succumb to the opposition. It is a vexing problem prevailing in almost all states of the country. A solution must be found for consumers to pay reasonable fees for the services they receive and all subsidies should be progressively reduced and ultimately phased out. If this is not done, the constraint on financial resources would continue and funds to upgrade water use infrastructure works and their efficient operation and maintenance would not become available. The efficiency and quality of service would continue to remain unsatisfactory as at present.

The distribution of financial resources between different development sectors is done on 'annual' and on 'five-year' basis through annual and five-year plans. The allocations are generally made on the basis of 'general perception' in an adhoc manner. No exercise is done to evaluate the comparative economic return per unit of investment on development sectors like power, industry, water, highways, etc. Priority needs to be given to development works on the basis of economic return they generate. Irrigation sector which contributes more handsomely than any other development sector to rural development, making drinking water (good quality) available, in making the state self-sufficient in food, in generating employment and in raising the economic status of farmer community, has received a raw deal. Upto the Fourth Five-Year Plan period, irrigation sector was accorded priority. Thereafter, the financial allocations to 'irrigation' sector as a percentage of total plan expenditure, have been declining rapidly. The allocation to irrigation sector which was 45 per cent during the First Five-Year Plan came down to 15 per cent in the Eighth Five-Year Plan as shown in Figure 8.6. The representations by the Irrigation Department, bringing out this position, and seeking higher allocations bring no response from the State Government.

Figure 8.6

Outlay on Irrigation Sector and IGNP as Percentage of Total State Plan Outlay

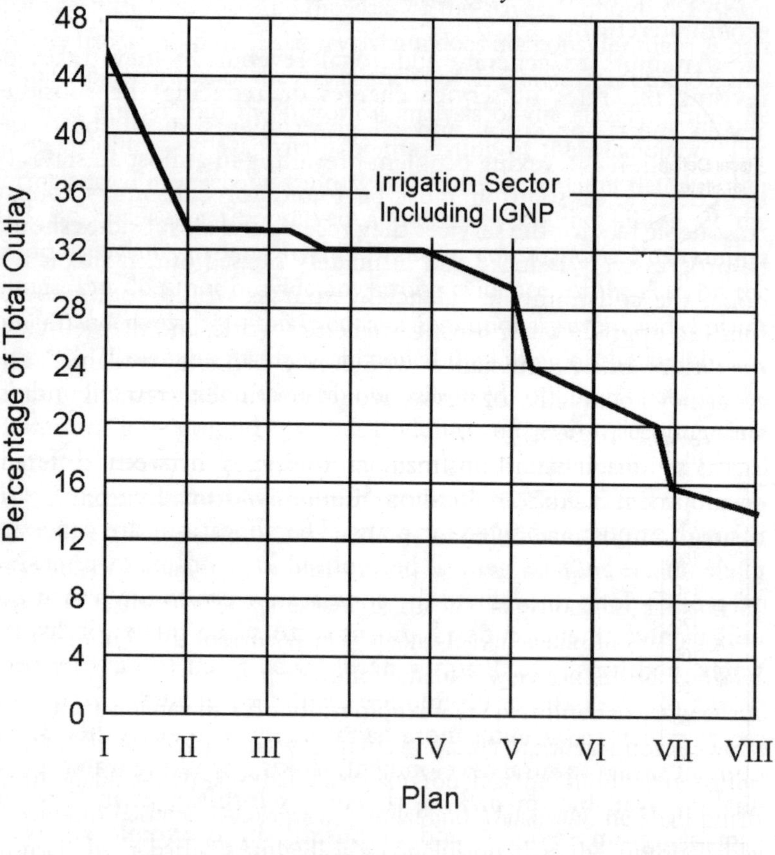

Unsatisfactory Maintenance and Management

There is urgent need of improving management and distribution of water to avoid wastage, improve reliability of water supply and ensuring equity in distribution. This would require proper upkeep and maintenance of irrigation works, involvement and active participation of water users in planning, implementation, operation and maintenance.

Most irrigation infrastructure works are poorly maintained. Some, because of continuous neglect, are in a hopeless condition. Water is wasted and farmers, particularly in tail reaches of canals,

Figure 8.7
The Vicious Circle

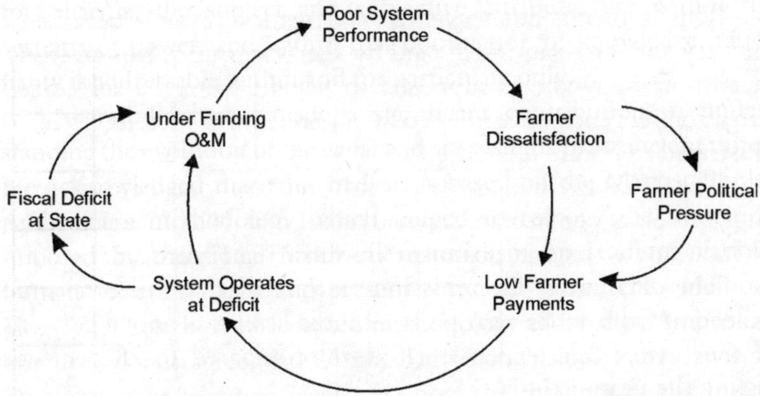

Figure 8.8
Temporal Behaviour of Direct Receipts (per ha) at 1988/89 Prices

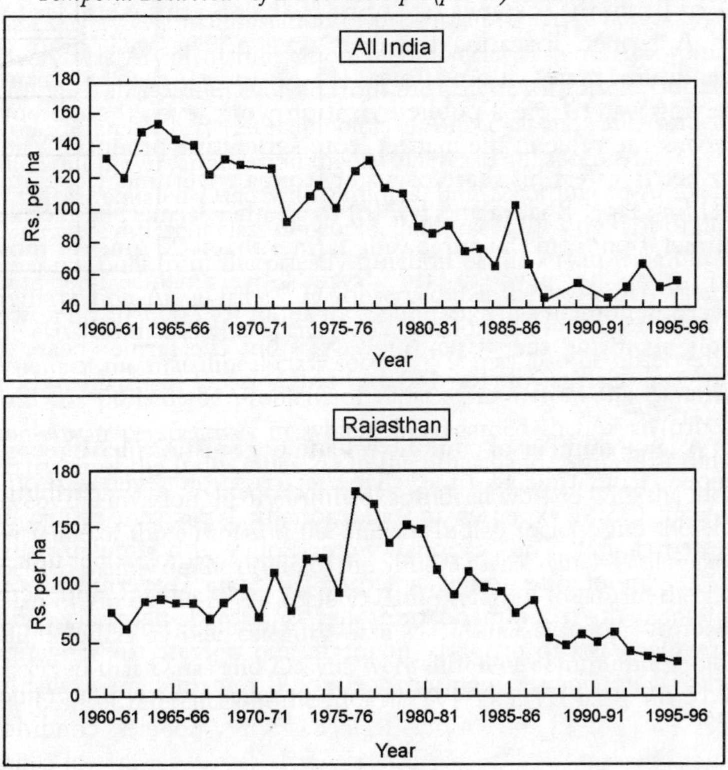

suffer badly. The primary reason of this is very inadequate allocation of funds for maintenance.

This in turn is due to very low irrigation revenue because of the very small irrigation fee paid by the Water user. It is an irony that the farmer who is supposed to benefit from the subsidized irrigation fee, ends up as the worst sufferer. It is a vicious circle, as illustrated in Figure 8.7.

The NCAER had made a study on temporal behaviour of direct receipts per hectare from irrigation at 1988-89 prices. The all-India and Rajasthan positions are shown in Figure 8.8.

The Planning Commission reported that the revenue collection from water charges at all India level was only 6 per cent of the total O&M expenses (1995-96) compared to 78 per cent during the years 1974-76.

The State Government of Rajasthan has recently revised the irrigation rates, bringing the overall average rate to about Rs 100/ha (from the previous rate of Rs 50/ha). Bu this is still far too low. A farmer, irrigating from his own private well, bears an expenditure many more times (15-20 times) than a farmer receiving water from a public irrigation work, even though both receive same value in the market from agriculture produce. When a farmer transfers his share of water from a government operated canal (in Gang, Bhakra and IGNP) to another farmer, he receives remuneration from the purchasing farmer that is 20 times or more than what he pays to the State Government. The State Government incurs an expenditure of about Rs 2,000/ha on power supply in lifting the water (in IGNP) but the farmer bears an additional charge of Rs 100/ha only, Is it Justifiable? Is it sustainable?

A large number of committees and commissions have recommended, from time to time, that the irrigation revenue should cover all O&M expenses and additionally 1 per cent return on invested capital. The National Water Policy also stipulates this. But, the moot question is whether the State Government can implement the recommendations and/or provide necessary funds on regular basis to properly maintain and operate the irrigation works. If not, better maintenance would remain a mere wishful hope.

Water Users Participation

The planning, implementation and operation of most water schemes come under the domain of government agencies. Laudable work has been done by these agencies in harnessing water resources and in operation and maintenance of major works like dams and large canals. But, the same cannot be said about maintenance of smaller canals and distribution of water. They are in poor shape. The government agencies believe they have no role to play below the outlet and do not know low water is distributed amongst farmers, except in the Gang-Bhakra and IGNP projects. This results in poor efficiency in water use and gross inequity in water distribution. Those in the head reaches of canals and the powerful ones benefit at the cost of others.

The minor tanks and small works like anicuts, etc., fall under the jurisdiction of the village panchayats. The position here is no better. Leaving aside a few exceptions, the village panchayats look for government aid for these works. They have done little to improve the efficiency of water use and in ensuring fair distribution of water.

It is now generally accepted that to remedy and improve the position, active participation of water users is necessary. Government departments alone cannot bring about the desired improvements. The village panchayats and the water users, by themselves may not be able to achieve much. The best chance is if they and government agencies work together. But this is easier said than done.

A beginning has been made. The Irrigation Act and Rules have been amended to accord legal authority and put responsibility on the Water Users Association. Rules for formation and role of the associations and working committees by election have been put in place. Training programmes are being organized. It is too early to gauge the success of these measures. But this certainly is a step in the right direction. To what extent it would succeed would depend upon the departments of the State Government, and the farmers themselves. Public representatives and NGOs can play a helpful role. The government should evolve and put in place a system to monitor and assist this programme to ensure its success. This would need untiring efforts.

Groundwater Recharge

The overexploitation of groundwater, and its deteriorating quality, are a matter of great concern. The groundwater use is largely under private sector. Because of the growing demand, it is difficult to put a curb on groundwater use. Government interference to control and regulate the groundwater use would create more problems than it would resolve. At the same time, inaction on the part of the government would make the situation from bad to worse. It seems that matters would continue as they are until the groundwater table goes down too deep to be economically exploitable and/or the groundwater quality deteriorates to the extent that it cannot be put to domestic or agriculture use. We are heading in this direction,

Recharge of groundwater can be a part solution. Water harvesting in the countryside and roof-top rainwater harvesting in urban areas can help. But these can play a limited role. Ultimately a cheek on growing demand (by control on population) and increasing surface water potential from external resources would be necessary.

Conjunctive Use of Water

In the command areas of major projects like the Chambal, the IGNP, the Gang and Bhakra, the recharge from irrigation is causing rise in the groundwater table. In some areas the water table has come up to the ground level causing water logging and salinity problem. In such areas, groundwater can be used conjunctively with the canal water. This will provide twin benefits. One, additional water resource to an extent of about 30 per cent (of canal water potential) would become available. Two, pumping out of groundwater can check rise in groundwater table and help in mitigating the problem of waterlogging.

There is, however, little conjunctive use of groundwater in the command areas. The supply of canal water comes quite cheap to the farmer. Use of groundwater involves costs in making wells and in pumping. This results in no use of groundwater. This anomalous position cannot change unless the cost of canal water to the farmer is made comparable to that of groundwater.

Small Schemes and Big Projects

There are many who believe that the country's water problems can he solved by opting for small water storage schemes, watershed management and the old traditional measures. It is stated that the large grandiose schemes of big dams, river links and large canals are not needed because they cause great ecological damage. The whole matter is argued choice between the 'small' and the 'big'.

The policy and practice at the national as well as at the state level is to promote both the small as well as large schemes. Several thousand small water schemes have been constructed in the state. Farmers and village panchayats are free to take up small works. Small storage reservoirs are constructed on small streams wherever feasible. But, to harness the water that flows in the large streams and rivers during the rainy season, large dams are necessary. Small and large dams play complementary roles. They cannot be substituted one for the other.

It can be shown that in terms of submergence area, loss of water by evaporation, construction, operation and maintenance costs per unit of stored water, large dams and projects are significantly cheaper than the small ones.

The opposition to large projects is based more on sentiments and nostalgia relating to yesteryears rather than on an objective evaluation of the factual position and present and future needs.

Conclusion

The growing population in Rajasthan is putting great pressure on the limited resource. There is a need to control population growth to check the growing demand. To improve the supply, all unutilized surface water potential in the state should be harnessed. But this would not suffice and the state has no option but to look towards maximum possible supplies from external sources. The inter-linking river project envisaging transfer of water from surplus to deficit river basins offers an opportunity that must be seized. The state should work towards this tirelessly.

Agriculture is the dominant user of water (80-90%). The prevailing water use practices are very inefficient. These must be improved by proper upkeep, maintenance and operation of water conveyance systems and adopting water saving methods (sprinkler and drip) of irrigation in the field.

The current low irrigation fees are unsustainable. Lack of revenue results in grossly inadequate maintenance. This in turn affects the quality of the service. The ultimate looser is the farmer, who is supposed to benefit from the state subsidy in the form of low irrigation rates.

Groundwater is already being over-exploited. We are progressing towards a dead-end. The state policy of extending power supply in rural areas for agriculture and making power available to farmers at subsidized rates is counter-productive.

Conjunctive use of water in the command areas of large projects, should be promoted. Water users' participation in planning, implementation and operation of all schemes is necessary. A start has been made in forming WUAs. Many difficulties would be encountered. Early failures can result in loss of confidence and cause a collapse of the scheme. The state should ensure success.

Opposition is often voiced against big dams and schemes. It is propagated that old traditional methods of water storage and small watershed schemes would serve us better. Undoubtedly, the old traditional storage methods and small schemes can play a useful role and should be promoted. But these alone cannot meet our present needs. Rajasthan's population one hundred years back was one-fifth of what it is today and two hundred years ago, it was perhaps only 20 lakh in all. Nostalgia about old things should not come in the way of pragmatic approach to solve present day problems.

Chapter 9

Training for Participatory Irrigation Management in India

Rakesh Hooja

In many countries, irrigation sector reforms and Participatory Irrigation Management (PIM) with increasing transfer of management responsibilities from government agencies to farmer or user organizations are gaining momentum. Mexico, Philippines, Colombia and Turkey are some of the well-known examples of transfer of irrigation systems through PIM processes. In India, even though PIM is being promoted for some time the scale and impact has been somewhat negligible.

The first shift in thinking on irrigation management can be traced to the multidisciplinary Command Area Development (CAD) programme in the 1970s, which envisaged optimization of irrigated agriculture in a manner that should improve the quality of life of the people within the Command Area, as opposed to the traditional approach of constructing canals thus creating irrigation potential and hoping that it would result into increased area under irrigation which in turn would lead to increased agricultural production. However, after more than three decades of efforts under CAD, what is disturbing in India is the failure of even the best CAD organization to evoke much farmers' participation except in some rare instances. This could be attributed to attention being focused on hydrological, engineering, agricultural and economic aspects. Especially in official documents it has in the past been rare to find described, let alone analyzed, the human side of the organization and operation of irrigation systems. The management of those who manage water, the procedures of allocation of water to groups or individuals, and the distribution of water within groups, there may have been almost as many instances of those omissions as there are reports on irrigation. Hence, efforts to involve farmers in water management are like groping in the dark due to insufficient knowledge and understanding (Hooja, 1995).

PIM was introduced on a pilot basis as part of the CAD programme in various Indian states. Then, in 1996, the Ninth Five-Year Plan Working Group on PIM of the Government of India (GOI) suggested that irrigation schemes covered under centrally-sponsored CAD Programme or under World Bank-aided projects be taken up first for introduction of PIM. It also suggested mass awareness campaigns and large-scale training of farmers and government functionaries and corresponding training capacities for PIM to be built up. However, in most of the cases, these efforts have had limited success.

The inevitability of reforms and the urgent need for farmer participation led to renewed impetus for PIM. Andhra Pradesh became the pioneering state in India to bring about overall institutional reforms in the irrigation sector at the state level in 1997 with a 'big bang' approach (Joshi and Hooja, 2000). The Andhra Pradesh Farmers Management of Irrigation Systems (APFMIS) Act 1997, became the model for reforms in Madhya Pradesh. Both Andhra Pradesh and Madhya Pradesh have witnessed two rounds of elections for Water Users Associations (WUAs). Some other states have opted for changes or replacement in existing irrigation legislation. In most of the states, however, the momentum for PIM has been slow.

The experience of the pilot projects on PIM under CAD, as well as the experiences of the various states, highlights the importance of training as an imperative component of any reform process, critical to the realization of the objectives of the programme. Through training, the farmers and the staff can be equipped to handle new responsibilities and to cope with changed job requirements. The limited success in many cases also suggests that training needs to be developed as an instrument of change in the reform process rather than as a support mechanism, which it is currently viewed as. Hence, the concept of training has to be broad-based and flexible for incorporating new challenges.

The training needs for PIM in the country would vary from state to state depending on the progress of PIM and the changes that are envisaged, but for the country as a whole they are definitely immense. The report of the Working Group on PIM programme for the Ninth Five-Year Plan of the Government of India estimated that nearly 30,000 farmers and 2000 officers would need to be trained during the plan period. These estimates were based on the proposal that 2000 pilot projects would be taken up during the plan period and that at least 15 farmers and one officer would be trained per project. Further, the Working Group suggested the need for increasing the capabilities of the present training institutions and the creation of proper training modules (GoI, 1996:47).

These estimates by the Working Group sound modest if compared by the magnitude of training man-days generated during the adoption of the new Irrigation Act in Andhra Pradesh. During March-April 1999 alone, 11 intensive four-day training programmes were conducted for the training of trainers to become the district level trainers. Five officers each from irrigation and CAD, revenue, and agricultural departments were trained as trainers for the districts. A "Trainers Hand Book" and a "Manual on Orientation in Training" were brought out for the guidance of the trainers. A special 'training kit' had been designed for training farmers with flip charts and a manual prepared in local language. During July-August 1999, the training programmes were conducted in all the districts at selected places for the training of 65,000 presidents and Territorial Committee (TC) members of Water Users Associations (WUAs). About 55,000 presidents and

TC members were actually trained during this period. Further, these trainees were given a manual on WUA activities and details of the provisions of the APFMIS Act with description of the rights and responsibilities of the WUAs.

Given the aim of training to function as an instrument of change and to help to shape the reform process, the issue of designing appropriate training programmes should be addressed so as to enhance effectiveness of the training programme. The design of the programme should clearly follow from the objectives of the training programme, and the follow up and feedback received from previous programmes as a measure of the achievement of the programme objectives. A clear definition of the objectives for any training programme forms the first essential step for designing a training programme. A clearly spelled out purpose for the training could include objectives such as:

- To develop a cadre of committed people through the training programme with the conceptual know-how about PIM and the willingness to propagate it.
- To reduce the prejudices and problems, through stimulating modified attitudes and behaviour suited to the adoption of new paradigms in PIM.
- To enable participants to acquire practical knowledge and skills for various definite PIM related purposes, for example, for creation and management of WUAs, or for the use of agriculture extension and irrigation extension efforts of the government to facilitate, propagate and stimulate PIM.

The objectives of the programme would depend on the type of trainee group and the specific tasks that the specific group is expected to handle in the PIM process.

This chapter adopts a view of training as an instrument of reform and stimulating changes in the existing systems and practices, thus removing it from the confines of the traditional, narrow outlook of training as a means to impart information and skills about particular aspects or subjects. Thus, in the context of this chapter, training is being viewed in its broadest sense to include mass awareness building and attitudinal change in addition to imparting skills and knowledge related to irrigation, agriculture, participation, group dynamics, financial accounting, etc. The PIM training would thus need to:

(a) create a favourable climate within the country in general for the propagation of PIM,
(b) prepare water users to undertake PIM through Water Users Associations of some sort or the other,
(c) enable the government departments and agencies related to water resources and irrigated agriculture to reorient themselves and adopt new roles, organizational structures, processes and ways of functioning such that they can play a facilitative role for furthering PIM and joint farmer-government agency partnerships, and
(d) equip lower level functionaries of government, employees of NGOs, and others to facilitate PIM.

As has been indicated in Hooja (2000), training for PIM is too important to be left only to trainers in places like WALMIs. Multidisciplinary teams of those knowledgeable about social interaction, irrigation, agriculture, groundwater, training, administrative coordination, rural development, extension services, etc., need to collaborate with government functionaries, farmers and NGOs, who are intimately involved in implementation of PIM for imparting training in PIM. Thus, the multidisciplinary mobile PIM teams being used for implementing PIM programmes and government staff, chairmen of WUAs and community organizers involved in the PIM pilot projects, should all be jointly used along with specialist trainers fully oriented about PIM to impart PIM training and orientation. The state level PIM cell responsible for policy making and implementation of PIM should directly coordinate such training. Thus, a PIM Training and Orientation Centre should be placed under the direct control of the state level PIM policy and implementation cell in the State Government Secretariat.

Sinha (2001) has suggested the following objectives of training in PIM:

- Increased participation of the farmers in every activity related to water management.
- Enabling farmers to identify the problems of the irrigation system to seek their own solutions and formulate and implement their own projects to set right the system.
- Rehabilitation of irrigation system especially canals, through participatory processes.

- Operation, maintenance and management of irrigation system by farmers.
- Equitable distribution of water among all irrigators.
- Additional resource generation for irrigation management activities.
- Increased involvement of women and youth at all levels of irrigation management.
- Reorientation of irrigation bureaucracy, field staff of line agencies and high-level policy makers including the government bureaucracy and politicians.
- Increased participation of NGOs, community organizers and voluntary organizations in PIM.
- Promote genuine NGOs, trainers and social activists who can facilitate the training objectives in PIM.
- Training of trainers in PIM.
- Environmental safeguards against waterlogging, salinity, alkalinity and water table recession.
- 'Social capital' building in the form of Water User Associations (WUAs).

Hooja (2000) goes on to add that PIM training should include:

- efforts at general orientation about PIM and attitudinal reorientation about PIM,
- knowledge and skills in improved water and agriculture management, i.e., technical matters,
- PIM-related procedural, administrative and organization issues, incentives available, sources of incentives and funding,
- lessons (about both successes and failures) from a variety of projects in India and abroad, and
- commercial matters and accounting systems (e.g., *bahikhata* system) which local farmers may be able to understand.

PIM training will be required not only for all farmers using surface irrigation, but also for all officers and staff of irrigation, agriculture and CAD departments (and perhaps even for groundwater department and district rural development agency or Zila Parishad staff).

Extensive use of mass media should be made. In addition to the Central Government generating and providing mass media packages, at the state level also video and audio cassettes, TV and radio programmes, newspaper articles, pamphlets, leaflets, posters have to be prepared and disseminated on a continuing basis. Schools and colleges could be tapped through debates and essays. Planning forums in colleges and the National Services Scheme could also be used to make people aware and orient them about PIM; its need, the problems in its implementations, as well as the likely benefits. Government agencies staff will also have to be convinced and motivated about the need for them to take a proactive approach to the propagation of PIM, as would policy makers and politicians.

This chapter aims to stimulate further thinking on the training needs for PIM in the country. Hence, it also identifies some policy issues related to PIM training at the national and state levels.

Identification of Trainees and Training Needs

Capacity building constitutes an imperative component of an irrigation reform process (or a PIM adoption process). It encompasses all the stakeholders in the PIM process, from creating a supportive policy and administrative environment to building capable and sustaining organizations at the farmer level. A complete PIM training programme for a state should include:

- Appreciation workshops for politicians, bureaucrats, senior level officers including the ministers (especially Irrigation Minister) Secretaries of the State Irrigation, Command Area Development, Agriculture and Groundwater Development Departments, Chief Engineers, etc.
- Appreciation workshops for all officials of the irrigation department and other concerned departments.
- Specific capacity building workshops for community organizing teams of governments section officers and field staff as well as other community organizers.
- Specific training for the office bearers of the farmer's organizations/Water Users Associations.
- Training for local NGOs.
- Training of trainers for each category of trainee group and for skill development in the different subjects of training programmes including social sciences, agriculture, agronomy, engineering, etc.

- Case study exercises where participants, experiences and systematic examination of case studies would help to clarify concepts, issues and problem areas amongst the participants and show the way forward.

The training programmes would have to include various departments such as Agriculture, Irrigation, Groundwater, Command Area Development and Rural Development. Further, district administrative staff including the Collectors, Zilla Pramukhs and Divisional Commissioners should be involved in the PIM process. Their involvement and support becomes essential in arranging for large-scale training programmes and awareness camps, dealing with extension efforts and aspects related to the collection of water fees and the various changes expected because of PIM. Farmers, members and the office bearers of Water Users Associations would need intensive training for handling new responsibilities and roles. Of special importance is training for trainers, researchers and scholars, evaluation experts, teachers in academic institutions. Civic society involvement and support needs to be developed by including NGOs, media personnel and the public at large with orientation training on PIM.

It is important to categorize these trainees into groups as per the roles they perform. The training and orientation needs of each category (or sub-category within it) may require different types and different modules of training. For example, Agriculture Department staff would require somewhat different training from Irrigation Department staff or from Command Area Development staff. Further, the categories of trainees would vary in different states, requiring the process of training needs identification to be based on detailed assessments at the state or regional level. However, a broad categorization is discussed below as an example:

Politicians

One of the major bottlenecks in the reform process is lack of political commitment to the reform agenda and lack of political leadership. The objective of having orientation programmes for the political leadership is to acquaint them with the developments in the field of PIM in various countries. These programmes would equip them with information on the process of irrigation reforms and the political implications thereof including possible benefits of

PIM for the members of the WUAs, for other farmers, for the community at large and for the government. They may thus come to realize that good water management can also mean good politics as well as development of the area and its peoples.

Senior Management

The attitudes of the senior management are critical to the success of the PIM programme. Attitudes may be difficult to measure but make the difference in outcomes. This category of trainees would benefit from an exposure to the best practices adopted by other countries, and the progress of PIM in other areas. It is important to identify various departments like the Irrigation, Agriculture, Rural Development, CAD, etc., as the training needs would be different from one group to another. For the Irrigation Department, it would be imperative to stress on attitudinal change towards farmers, and for a movement from complete irrigation agency authority over management to joint management of the irrigation systems and the increasing role of the farmer associations. For the Agriculture Department personnel, they can be expected to have been more in contact with the farmers in their work than the Irrigation Department. Hence, the training would need to emphasize more on the technical aspects of water management and the problems of organization of groups in rural areas.

Middle Management

As key implementers of the reform process, they need to be convinced of the inevitability of the reforms and the training should focus on removing the fears and problems associated with uncertainty in the face of change. Again, there would be several categories within this group from different government departments or even specialities within them for example, the extension staff of the Agriculture Department may be sub-categorized not only for orientation training but also as trainers for the farmer associations.

Lower Management and Field Staff

These are functionaries who are directly involved as community organizers. They need to be informed about the thinking on PIM among the political leadership, and the senior management and how things would be changing with the implementation of the reforms. Again, attitudinal change would be extremely important, as they have to develop new relationships with the farmers and the WUAs. They would also require skills related to effective supervision for tasks that may be performed jointly by the WUAs and the Irrigation Department, or CAD and Agriculture Departments. Sub-groups within this category may be identified on the basis of project boundaries, regional differences or different systems depending on the nature of reforms and the progress of the reforms within the states. Also, for this category, training should be an ongoing process related to field implementation and not a one-time activity. Follow-up and post training seminars, or repeated follow-up training, would ensure that the participants have time to think and adapt to the changes and have access to regular upgradation of knowledge and skills as well as continuous impetus for attitudinal change. Agriculture extension, irrigation extension, on field water application, rotational running of canals and water budgeting, undertaking of repair and construction works of a minor nature are amongst the subjects whose knowledge would need to be imparted.

Farmers

Training for the farmers is critical not only for the development of sustainable WUAs, but to bring about basic acceptance of the concept of PIM. Farmers should be able to discern that there are sufficient gains from assuming responsibilities that were previously thought to be those of the state. Because there are substantial monetary and non-monetary costs to the farmers for expanded WUA (Water Users Association) activities, unless the anticipated benefits are also substantial, farmers would clearly choose not to get involved (Meinzen-Dick *et al.*, 1995). Further, effective leadership among the community is a key factor for success of the reform process.

Training of the users-cum-office bearers of the WUAs would need to incorporate several skills that are new for the farmers such as account management and book keeping, record maintenance or group functioning. This is the group that would require maximum training covering each aspect of training, viz., knowledge, skills and attitudes. They have to be encouraged to develop vibrant level local organizations through transparent management, group functioning and effective conflict resolution mechanisms. It is the development of a new rural leadership at the minor canal/distributory canal command level, which is likely to face problems both during its infancy as well as later. Unless sufficient time and resources are invested in building capacity at this level, it could become a threat to the entire reform process, as the farming community has to be prepared to take up the new responsibilities and to handle new challenges. The farmers would also need to be able to constantly update their irrigation and agriculture related knowledge and skills.

NGOs

NGOs can provide substantial back-up support in the form of training institutions and documenters of the reform process. At the implementation level, the NGOs could also function as community organizers. They could be involved as trainers to the irrigation, CAD and agriculture departments and of the farmers. They could facilitate the development of training material especially in local languages. Process documentation is another aspect where NGOs can function as effective observers and documenters. Hence, training of trainers would be useful for them, as well as equipping them with necessary training facilities.

Trainers

The training of trainers is a crucial activity in any training agenda for the state. There would again be several categories of trainers, who can train different categories of trainees and cover different subjects. Given the magnitude of training programmes that would have to be organized at the regional levels for imparting training to all the personnel concerned, trainers in significant numbers would have to be available for undertaking these training programmes. Hence, it would be imperative to organize several training of trainers to identify and train resource persons for the programme.

Implementers from various departments from different states and sub-regions that have initiated reforms, especially Andhra Pradesh, should be involved for this activity, as they would be in the position to share actual experiences for the reforms in their states.

Devising Training Modules

Given these different trainee categories with different training needs, it is important to devise training modules suited to their respective needs. David Groenfeldt and Peter Sun (Groenfeldt) briefly list them out:

- *For policy makers:* Awareness training to become familiar with the concept of PIM and the supporting policies it requires.
- *For mid-level officials:* Awareness training with detailed information about best practices from countries that have adopted PIM. It would be useful for them to tour successful areas in different states in India.
- *For agency staff:* Details on organizing and communication skills, legal and financial aspects of establishing WUAs, etc.
- *For WUA leaders:* Organizational and financial management.

This list, however, does not include the attitudinal and behavioural aspects that are critical components. The challenge is to train engineers, who have earlier been trained to think only in terms of number of hectares in command, storage capacity, length of canal, etc. One of the most important things, particularly for Irrigation Department officers, is how to start the PIM process and how to approach the farmers (Gupta and Srivastava, 1999). There is not only a change of attitude involved but also a reversal of roles. They have to build rapport with the farmer community, and to facilitate their efforts, patiently and persistently. Each individual staff member will have to understand his existing behaviour, attitude, role and perception, empowerment, and change, all along with working styles.

Annexure I gives some details on the topics that could be covered for training of different categories of trainees. The suggested topics are indicative and have borrowed from many sources. Each state, or sponsor of trainings, or training institution would have to design its own specific course for each type of training objective and for each type of trainee category.

Categorizing Trainees

We may suggest categorization of trainees for PIM as follows: (1) politicians (at central, state and various sub-state/local self-government/Panchayati Raj levels), (2) farmers, (3) WUA personnel (elected office bearers as well as WUA employees), (4) NGOs, (5) community organizers, (6) senior officers (in central and state governments as well as district collectors), (7) middle level officers (with generalists and specialists from agriculture, irrigation, CAD, groundwater, rural development, panchayats, etc., departments and from the district administration), (8) junior level officers (especially those related to irrigation CAD and agriculture and such other departments with whom WUAs are likely to have regular interaction), (9) trainers (from WALMIs/IMTIs/ WALAMTARIs or other training institutes related to agriculture, rural development, panchayats, etc., or for trainers picked from amongst government departments implementing PIM), (10) mass media or from amongst WUA chairpersons or from NGOs, etc., (11) professionals, and (12) the general public.

While the contents of the training programme should encompass various aspects, what needs to be investigated is whether such courses already exist, or, whether existing course could be modified, or would totally new courses need to be planned for. What adds to the complexities of the various categories and their different training needs, is that the needs would also depend on the stage of the reform process in a particular state or region, and that training is likely to be required as an ongoing process. The next section details some of these complexities and requirements for the training programmes.

Training: Aspects and Methods

Training for irrigation reforms is an extremely complex task of balancing diverse categories of trainees with experience in diverse hydrological settings and different historical water management regimes and the training needs of the region or irrigation projects. This necessitates the formulation of a detailed training plan at regional levels. Different states would have different focus depending on the following:

- *Nature of the reforms:* states that have new legislation, or those where existing legislation has been modified, or where no legislative changes have been affected; and
- *Stage of the reform process:* states that are just initiating the reforms and those that are now into second generation problems of the reform process.

Further, training should have to account for regional differences in irrigation and agriculture practices and the historical mindsets associated with each system. For example, North India *warabandi* states may require different type of orientation and training as the farmers and management both are against changes in the rotational water supply system and favour the functioning of the traditional *warabandi*. In states where *shejpali* is practised, there is greater scope of involvement of the farmers in management of water supply below the outlet and correspondingly the requirements for training would be different. Adding to the complexity is the diversity in the scales of the irrigation systems – major, medium, minor, tanks, etc. These complexities would decide the objectives of the training programmes,

For specific training programmes, the aspects of the training programme need to be based on a comprehensive exercise on the training needs assessment for the target group and the stated objectives of the training programme.

Aspects of Training

The different aspects that training should cover three broad categories are as under:

- Knowledge and information for planning and appreciation of the programme.
- Practical guidelines and skills for effective implementation.
- Attitudinal change.

Attitudinal and behavioural change is an important component of the training for PIM. Most of the current training programmes encompass only knowledge or skills and do not adequately address the attitudinal components. The importance and need for attitudinal change was brought out effectively through a study that highlights that the irrigation bureaucracy is hardly aware of the variance between the department's own perception of its performance and the image as reflected in farmer's perception. These differences were effectively demonstrated by an exercise conducted separately with the farmers and the department officers in Gujarat state. The results are presented in the Table 9.1:

Table 9.1
Exercise in Perception

Image factor	*Agency's self-image**	*Perception of the farmers**
Planning for irrigation	8	1
Adequate water distribution	10	1
Availability of water when required	10	5
Maintenance of distribution/sub-minor	8	1
Distribution of water according to turn	8	1
Recovery of water charges	6	10
Satisfaction of tail-enders	8	2

* Scores out of 10

Source: Anil Shah in a lecture delivered at International Training Programme organized by Indian PIM at Hyderabad January 2001.

Training Methods

Training methods vary with the aspect of training. There is a need to develop new methods of training. Given that the average age of the trainees would be fairly high, the methods used have to suit the age and the experiences of the trainees participating in the programme. It is now fairly well established that the usual classroom teacher-taught type of training is not going to have any lasting impact. Further, training material has to be developed to suit the methods employed for training.

Some of the considerations in choice of the training method (workshop), include the objectives of the training and the learner's needs. As was stated earlier, the objectives for training different categories of trainers is likely to be different – the training of farmers and WUA office bearers could have specific objectives like building skills for record keeping and account management, while that of the politicians may be an orientation on PIM. Language of training may also differ according to category of trainee and type of training.

Further, the needs of the organization and the resource constraints – both time available and cost of training – has to borne in mind in the design of the training programmes and choice of methods. Exposure visits may require greater budget and training duration than classroom training. One of the limiting factors in the choice of training is often the non-availability of instructional material or poor quality of material. Many of the training institutions have been using outdated training materials and well-circulated references. There is an urgent need to design suitable training modules, with acceptable standards and to identify material that could be used for developing session-wise training materials. This is especially relevant since individual training programmes may not have the time or the financial resources to support development of training materials for each training programme.

Some of the training approaches usually adopted and suggested are briefly summarized below:

Knowledge and Skills

For this component of the training programmes, it is an oft-expressed view that implementers should be involved in training as they can share from experience the process and problems in reform management. Also, the acceptability of the implementers is higher to the trainees, especially to the engineers of the Irrigation Department compared to NGO trainers who are essentially viewed as outsiders having little knowledge of what the job of the engineers or system management demands.

Where training is needed at the field level, for agency field operations staff or farmers, experience suggests that on-the-job or learning-while-doing training and peer training (such as farmer-to-farmer) often works better than theoretical training. Exposure visits to demonstrate what the situation after PIM reforms is in other areas is also an extremely powerful tool.

Another debate on the training involves different views on the acceptability of traditional knowledge and procedures as against promotion of modem, 'scientific' know-how. There is a view that knowledge of engineering concepts is essential for maintenance and operation of the irrigation system and modern knowledge is required also for other activities such as book keeping and financial management. Hence, training involves an effort to transfer these methods to the farming community without an effort to find out what methods they would adopt to achieve the same outcomes. For example, the rural communities could have their own accounting system or *bahikhatas* which could perhaps be used for the purposes of the WUAs as well. This would not only reduce the pressure on the new office bearers of the WUAs to learn new skills, it would enhance transparency among the community as most of them would understand their traditional ways of doing things while new knowledge remains confined to a few. However, in absence of an open communication and an openness to discuss ways and means, developers of the training material remain unaware of these alternate ways of viewing and doing things. Hence, in developing training material and methods, an effort could be made to involve the trainee group and have a participatory development of the training. This kind of participatory training as compared to prescriptive training would enhance the effectiveness of the programmes and go a long way in making training an instrument of change.

Attitudinal Change

For effective attitudinal changes, which are lasting beyond the training duration, it is suggested that training methods should include:

- small, informal, non-threatening, open-minded discussion groups with persons who are important and respect the participants
- role play enables participants to experience new attitudes
- situation replays
- interactive video tapes – feedback
- simulation, and
- preparation and discussion of case studies.

Feedback from the international training course on 'Capacity Building for Participatory Irrigation Management', held at Bari, Italy, September 11-20, 2000, suggests that it is useful to incorporate exercises that stimulate the development of region/project specific implementable plans by trainees and discuss these. Follow-up of the actual implementation of these plans after training has also been suggested as a mechanism to explore the usefulness of the training and to ensure continued support to trainees who are trying to utilize the training in their jobs.

Assessment of Training Programmes and Institutions

Training is Low Priority

Training remains a low priority activity in the irrigation sector. Most of the training is mandated by donor agencies. This is also reflected in poor attendance and very often partial attendance on the part of the trainees, thus resulting in the failure to achieve the objectives of the training programme and also adversely affecting the morale of the trainers.

Training Needs Assessment

There is no assessment of training needs in terms of qualitative and quantitative requirements. There are two essential steps. First, there is a need for an attitudinal survey of farmers, and Irrigation Department staff as well as other categories of potential trainees. Second, there is a need to conduct a Training Needs Analysis, because as regards PIM training, this being a relatively new area, the needs are enormous and their nature varied. This would also help the training institutions understand the requirements and the inputs that need to be build up.

Training Material

Training programmes are handicapped by availability of limited training material, especially in the local language. Efforts at developing methodology and curriculum are ad hoc, and often routine, with the result that both contact and style are pedantic and outdated. At best, training institutions evolve new training designs based on limited experiences and resources. Only a limited amount of feedback is obtained from participants, which could otherwise be useful in updating and improving the training materials. While this is true for written texts, audio-visual material is very often not available at all.

Structural Problems

There are many structural problems generally associated with training. The location of the training is a major factor affecting attendance, as the key officials involved in field projects cannot afford to be away from their station for a long time. Hence, key staff members skip training (Oza, 1997). However, due to practicalities, training is organized at the training institutes away from field sites.

Lack of Proper Planning of Training Course

In many of the training courses including those organized by some Water and Land Management Institutes (WALMIs) and by Water and Power Consultants Ltd. (WAPCOS), a comprehensive training programme is not prepared by the core faculty of the training institutes but major emphasis is instead given on guest faculty. The idea of inviting guest faculty is good so as to supplement with practical aspects and field experiences in motivation, organizing farmers and sharing of achievements. But such guest lectures are often not arranged in sequential order, nor is it ensured that a common approach is being followed and accordingly there is much duplication and even omission of important topics as the knowledge or the specialization of the guest faculty available influences the topics included in the training schedule. The organizers should prepare a schedule course outline and detailed lesson plans in advance. They should ensure that the course is well structured and all the required topics and themes are covered in a proper sequence and indicate specific topics and lesson plans to the guest speakers, after ensuring that no gaps are being left in the course.

Failure to Use Evaluation of Training to Improve Subsequent Courses

Individual training programmes are sometimes evaluated at the end of course in a casual manner to provide an evaluation sheet got filled by the trainee to the course sponsors or funding agency. Seldom is evaluation done to actually find out the trainees perceptions of what the course achieved, or to use the evaluation and feedback to improve subsequent courses. Impact assessment of the entire training effort in a particular state or a particular training institute via-a-vis objectives has not been done so far.

Limitations of Existing Training Institutions

Lack of Trainers

Most of the training on PIM is handled by the WALMIs and WAPCOS, the trainers being Irrigation Department engineers. They are deputed to the training institutes from the parent department, usually unwillingly. In addition to the lack of motivation for imparting quality training, they are themselves in need of training. They may lack exposure to multidisciplinary fields and the dynamics of socio-economic aspects and organizational methodology. Further, knowing that they are likely to be transferred back to the parent department, there is little motivation for developing skills required for a good trainer.

There is little effort to involve implementers with experiences who could contribute to the quality of the training programmes. External faculty is limited by the funds or the convenience as regards their availability.

Limited Funds

Financial resources usually are limited to the amounts required for classroom training sessions with little support for exposure visits, development of new training materials, etc. Further, there is little support for any kind of post-training follow-up activity.

Lack of Post-Training Follow-up

Training by and large remains a one-time activity. The stress is on meeting the stated capacity of the training programmes, rather than on proper participants' selection. Further, training is deemed completed after the schedule topics/lectures are over with very little emphasis on post training follow-up and use of new knowledge and skills. WAPCOS conducts training for officers from various states. They have no idea who comes for training and where they go after the course (Gupta and Srivastava, 1999). As a result, much of the training conducted has failed to create the desired impact, in turn reinforcing the already poor perception of the importance of training activities.

Choice of Training Programmes and Clientele

There is little emphasis on training of trainers. Farmers are largely not much involved in any aspect. Workshops and conferences at the national, regional and project levels are often a forum for exchange of ideas and experiences, but are restricted in their impact due to attendance by the same officials, NGOs and consultants/experts. Very often this leads to little value addition, with repetitive examples and discussions. Exposure visits are limited to some officers and a few farmers or WUA chairpersons. Further dissemination of experiences of conferences, workshops and exposure visits remain limited.

Challenges for Training on PIM

The sheer magnitude, diversity, costs and benefits of Participatory Irrigation Management (PIM) in India suggest the immensity of training challenge due to the following reasons (Sinha, 2001).

- Irrigation not only requires much larger capital input than most other agricultural methods but also the largest input of trained and experienced human resources. The total manpower from irrigation, agriculture, command area development and land development organizations directly involved in irrigated agriculture in India, is probably the largest in the world. Besides these agency officials, nearly two-third of the country's workforce is still dependent on agriculture for livelihood. Therefore, the population size of the target group of PIM training is stupendous.
- The diversity of irrigation system ranging from small private wells, shallow and deep community tubewells, community tanks, small embankments *(bundhies)* to some of the most complex large-scale reservoir fed/run-of-the river canal systems further add to the training challenge,
- On an average, water use efficiency in the existing irrigation projects is put at 40 per cent which means that there is tremendous scope for water savings with introduction of improved irrigation methods, techniques and careful selection of less water intensive crops.
- The share of agriculture in total water use in India is estimated to come down from the present level of 83 per cent to the

level of 73 per cent in the year 2025 due to competitive demands on water use from other sectors, particularly energy, domestic and industrial use. In view of such compulsions, the optimal use of water in agriculture sector will occupy centre stage in the PIM training strategies.

- Irrigation issues are intertwined with many sectors like rural development, agriculture, surface water, revenue, cooperatives, forestry, environment, energy, marketing, post harvest processing, banking, transport, etc. As a consequence, PIM training strategies have to be truly multi-sectored in character.
- Regional differences have to be given due consideration. Problems, profiles, and thereby strategies will exhibit fundamental differences from region to region. Different sets of interventions and customized approach are required for different agro-climatic regions.
- Language barriers between different regions of the country has to be acknowledged and education and training is to be conducted in regional languages.
- Training institutes and organizations have neither the inclinations nor the resources to cater for such complex set of training needs.
- In the absence of focused government policies for the same, the necessary coordination between line departments and various governmental and non-governmental training agencies may be difficult to achieve.

WALMIs as Training Institutes

Even after more than 20 years of existence, the WALMIs have not been able to emerge as significant knowledge institutions even for Irrigation Departments of respective states, leave alone the water resources sector as a whole. As the report of the Small Study Group (SSG, 1999) states: "Overall WALMI has been a great opportunity lost. WALMI has a long way to go before it becomes much of an institution, leave alone an institution of excellence."

The top political and administrative leadership at state level has not taken much initiative to find out why WALMIs have not delivered despite the fact that they were all born with silver spoons in their mouth. The Government of Gujarat was the first to worry about this issue. As a result in October 1998, it appointed a small

Study Group to study the working of the Gujarat WALMI. The mandate of the study was to assess the performance of WALMI, Gujarat, and to understand the reasons that account for the gap between its performance and potential and finally, to recommend to the government a course of action to transform WALMI into a high performing institution. The Study Group submitted its report to the Gujarat Government in March 1999.

WALMIs are subject to limitations. The report points out that WALMI remains a mediocre institution because of the following reasons:

- Its vision is limited being a captive institution of the irrigation department.
- No comprehensive view of water resources sectors.
- Approach and staff narrowly oriented towards agriculture engineering and agriculture extension-hardly any competence in social sciences, behavioural sciences, public systems, and management organization theory.

Further, during discussions with various officials of the WALMIs, other problems emerged which include:

- Poor attendance in the training programmes.
- Faculty is largely drawn from the irrigation department (and to some extent agriculture department) on deputation. This often results in short tenures. Further, postings at the training institutes are considered inferior to other postings and are largely 'unwanted'. Hence, there is an accompanying lack of motivated staff in the training institutes.

It is quite clear that the WALMIs and IMTIs in the country need major reorientation and capacity building if they are to become effective instruments of PIM propagation as well as PIM training. (An insider's personal view about capacity building of WALMIs is given in Sinha, (2001). "Education, Training and Other Capacity Building Measures in Water Sector for 21st Century." National Seminar 1-2 March 2001. Technical Papers, UPWALMI).

Measures for Building Capacity for the Training Institutions

There have been some initiatives from the government to identify shortcomings in the current training programmes and the corrective actions that are required. The Gujarat Government appointed the small Study Group for training in PIM with Anil Shah as its chairman. The Study Group concentrated on the training needs of Gujarat state and reported in November 1999 recommending: (a) give priority to training to field staff, (b) training should be in two modules (i) conceptual understanding of why PIM, the process of promoting WUAs and changing role of field staff from implementers to facilitators, and (ii) issues in working of WUAs and the actual role of field staff, and (c) of the 4,000 field staff, 50 per cent may be trained in next three years through the WALMI and three existing NGOs.

A training and awareness building strategy specifically aimed at propagating PIM needs to be worked out for each state of India. Some measures for building capacity including for training institutions are suggested below:

Involvement with State Policy Decision-making

Training is often viewed as an isolated, one-time activity with little impact on or relevance to the state's decision-making system. This perception needs to be altered and an awareness generated about the relevance of training in the policy making process. This could be achieved through exposure workshops of senior management officials as well as awareness camps for creating awareness about the relevance of training.

There has to be supporting policy change in the Central and state governments backing a comprehensive, need-based, demand-driven training programme. It should be promoted on the basis of detailed analysis of training needs in a state, based on projects or scale of irrigation, and careful identification of the trainees. The process could be initiated with the identification of priority states. These include those states that have passed new legislation for PIM and there is need to identify training programmes that have to be conducted before, during and after the enactment of the law following the examples of Andhra Pradesh. For areas where PIM has been under implementation for some time the training needs and packages would require to be different. Secondly, identification of the trainees has to be rigorous and need based, supported by evaluation of the impact of training through examining the post-training activities of the trainee.

Evaluation of Training Programmes

There should be mechanism designed to evaluate PIM training, both in terms of impact of a particular training programme, and in terms of the impact of the overall PIM training and orientation efforts in a particular area or the country as a whole.

ToT Programmes

Conducting more ToT (Training of Trainers) programmes based on needs analysis. Expert groups should be set up for development of curriculum and methodology. This process should also be participatory, especially incorporating the feedback from past trainees and from persons who have been course coordinators. Curriculum and methodology should be developed for each different category of trainees through PRA techniques with the assistance of consultants and local NGOs.

Physical Equipment and Infrastructure

Training institutions should be equipped with sufficient infrastructure and training materials, especially audio-visual aids, etc.

Exposure Visits

This is an important activity for all levels of trainees. Exposure visits to other areas for interaction with counterparts is considered an effective way of exchange of ideas and information.

Networking

While this idea is brought up at various forums, there is little progress in reality. During the meeting of the Committee for Training on Participatory Irrigation Management constituted by Ministry of Water Resources (MoWR) in 1999 (which unfortunately did not finalize its report or recommendations), the idea for creating a network for institutions promoting PIM in the country was mooted. The SPWD (Society for Promotion of Watershed Development) concept of a 'training grid' as has been considered for watershed development was also discussed. However, this had not achieved success due to lack of clarity on the role of the coordination agency (SPWD), or the incentives or disincentives for various institutions to become a part of such a network. This threat faces any proposed network for PIM training institution as well. Hence, initiatives need to begin with a reasonable consensus about the need for a network, the value addition for each participant and the incentives/disincentives in being participants to the network.

Clientele Group

The training programmes need to gradually be made more demand-oriented. However, for quite sometime most of the training would have to continue to be government or donor driven as training would be a major instrument of propagating PIM amongst those who have not realized the importance of PIM or those who need help with the operational details related to PIM.

Trainee Groups

The training imparted should be functionally relevant and this could be ensured through appropriate trainee groups, with similar needs, past experience and expected follow-up activity, rather than clubbing trainees from diverse backgrounds. However,, some horizontally integrated courses would also be required particularly where attitudes are to be changed. Some critical aspects of training that make a difference to the quality of training need to be emphasized. These are:

- Training is not a one-time input, but a process of continued interaction.
- Training programme should be oriented towards the needs of the client group and the objectives of the particular training programme.
- Networking between different institutes for exchange of ideas and other resources.
- Financial support should be in-built in the system for each activity. Support needs to be provided not only for the training component but also for field support during project implementation.
- Standardized training material should be prepared with maximum consultation and then adapted for the specific purpose of each particular course.

Recommendations

Broad View of Training of PIM

Training has to be seen not only as a means to impart knowledge and skills or to bring about awareness and attitudinal change, but as a means of propagation of PIM and speeding up reform in the water sector.

Resources for Promotion of PIM

The Government of India and the State Government as well as donor agencies would need to liberally provide resources for various types of awareness building and training courses related to PIM. The Government of India should share training costs with State Government or NGOs on 90:10 sharing basis and also fund WALMIs, etc., on a 100 per cent basis for national or regional programmes of PIM training.

Finalization of Training Modules

PIM awareness and training strategies need to be prepared separately for each state. This should be followed by preparation of a perspective plan for PIM training for each state. Training modules for different training objectives, different categories of trainees, and for areas at different levels or stages of implementation of PIM, would need to be developed.

Development of Training Material

There is need for a documentation centre/clearing house of training material produced all over the country and abroad. The variety of training courses and corresponding training material required for PIM is likely to be vast.

Creation of a Network of Training Institutes

A network of various institutes providing PIM-related training could be setup as a centralized source of information about forthcoming training programmes, and for evolving a common database for feedback and upgradation of the training programmes.

Creation of a Network of Training and Implementing Agencies

Networking between training institutes and implementing agencies who would need to use the training facilities would be useful in ensuring that the training is found useful and relevant to the client group.

Creation of a Network of Funding Agencies

Networking between the Central Government, State Governments and important agencies like World Bank, Ford Foundation, NABARD, etc., could further the training and orientation efforts so as to propagate PIM. This could lead to the development of comprehensive, state-wise plans and provision of funds and resources for training and identification of the trainees and their needs as per the state's overall objectives.

Training and Orientation is too important to be left only to training institutions

Policy makers, PIM implementers, WUA chairpersons NGOs, trainers all need to be jointly involved along with trainers at training institutions in planning and imparting PIM training. Their training skills would need to be enhanced and nurtured.

Need for Clubbing National, State and Regional Training with Local Training

From the point of view of sharing of experiences national or multi-state training would be desirable, however state or local level training would also be necessary to go into the details of PIM processes in particular regions or projects.

Annual Training Needs Assessment for Each State followed by Annual Training Plan

Keeping in view the objectives of the training, categories of trainees, state level of acceptance and spread of PIM, etc., and the state perspective plan for PIM training, each state should conduct once a year a training needs assessment specific to the state. Similarly, the training facilities and resources available to the state should be assessed. This should be followed by a detailed Annual

PIM Training Plan being formulated for each state. This plan should then be implemented efficiently.

Careful Planning of Each Course

Each individual course needs to be well planned, not based on easily available guest faculty or available training material, but on the objectives of the training, category of trainees, expectations from course, etc. Course coordinator's role is also vital.

Assessing Impact of Training

Not only should each training programme be evaluated, but the impact of the entire training programme in each state should be assessed once in six months.

Conclusion: Suggested Strategies for PIM Training

To recapitulate and sum up, the proposed PIM training strategy is being summarized below:

As regards policies and strategies for PIM training, the Central Government need to prepare a brief policy/strategy statement for PIM training exhorting the State Governments, water professionals, training institutes and other related people to give priority to PIM-related orientation and training and suggesting modalities for such efforts. The Central Government would also need to provide technical and financial support to various states for conducting PIM training. There is need to constitute a National Support Group on PIM to facilitate the Central/State Governments and other agencies involved in PIM training. Similarly, the government needs to sponsor the creation of a National Documentation Centre/Information Clearing House on training material and other documents related to PIM. Numerous national and regional level workshops on strategies and policies for PIM training need to be organized along with some national level training courses to enable inter-state interactions. Similarly, there is need for organizing research conferences on PIM which would, amongst other things, generate training material on PIM and provide feedback for policy initiatives and improved implementational strategies for PIM. Government should also facilitate the creation of a network of WALMIs and other training institutes across the country which are involved in PIM training. Networking between training facilities and agencies which can provide funds and resources for training as also with PIM implementers would also need to be evolved.

The Government of India also needs to facilitate the creation of model training manuals (which each state could then adopt to suit its own specific needs) and earmark funds for PIM training in various existing Government of India schemes whereby training and capacity building funds could be provided to various State Governments, training institutes and NGOs, for performing the role of a National Support Group and of a National Documentation Centre. For instance, the State Governments could be provided 80 per cent assistance for conducting a specific number of PIM trainings, for appointing a state level PIM Training Coordinator and for preparing PIM training material. The National Support Group and the National Documentation Centre could be provided 100 per cent assistance for PIM training courses and other related work including preparing standard PIM training material and suggesting strategies for more effective PIM training.

At the state level, both the Perspective Plan on PIM Training Capacity Building and an annual PIM training plan would need to be prepared, following a training needs assessment for each state which would encompass not merely the annual WALMI plans but all aspects of PIM training.

An agency (WALMI, an NGO, or a special cell in a government department) would need to be identified to act as the state level support group and documentation centre for PIM training and orientation. The State Governments would need to set apart earmarked funds for PIM training

As has been indicated elsewhere, on many occasions (see for example L.K. Joshi and Rakesh Hooja (eds.), *Participatory Irrigation Management: Paradigm for 21st Century*, by this author, training for PIM is too important to be left only to trainers in places like WALMIs. This is not to decry the useful role of WALMIs but to point out that multidisciplinary teams of those knowledgeable about social interaction, irrigation, agriculture, groundwater, rural development, extension services, training, administrative coordination and the like need to collaborate with government functionaries, farmers and NGOs who are intimately involved in implementation of PIM for imparting training in PIM. Thus, the multidisciplinary mobile PIM propagation teams being used for implementing PIM programmes, and the government staff, chairpersons of WUAs and community organizers involved in PIM projects, should be jointly used along with the specialist trainers in training institutions to impart PIM training and orientation. The state level PIM Cell responsible for policy making and implementation of PIM should directly coordinate such training.

A multi-agency/multi-disciplinary state level PIM Training Committee should be set up in each state having direct linkage with the state level PIM Cell overseeing policy making and implementation in the field. This committee should help prepare and review the Annual State PIM Training Plan, suggest activities to be undertaken, suggest aspects on which training manuals and model training modules should be developed, obtain feedback about training conducted, evaluate the impact of the training courses and suggest modifications, if required, in the approaches/strategies being followed, as also refinements in the subsequent years training plan. This committee should also be able to suggest as to how to attract funds from various sources for PIM training. It should also have links with all concerned State Government departments, training institutes (specially those related to agriculture/irrigation and CAD/rural development sectors), NGOs working on PIM, some of the more active WUAs, irrigation projects where PIM has started picking up, and also with the National Support Group and National Level Documentation Centre for Training on PIM.

State level PIM training and orientation manuals would need to be prepared with due attention being paid to different requirements of different regions of the state and to those WUMs which have just being created, as well as older WUAs facing second generation problems.

Formal training would need to be coupled with non-formal PIM orientation programmes and appropriate use of mass media channels. Extensive use of mass media should be made. In addition to the Central Government generating and providing mass media packages, at the state level also video and audio cassettes, TV and radio programmes, newspaper articles, pamphlets, leaflets, posters need to be prepared and disseminated on a continuing basis. Schools and colleges and especially Planning Forums in colleges and the National Service Scheme could also be used to make, not only students, but also all people, aware about PIM – its need, the problems in its implementation, as well as the likely benefits. Government agencies staff especially in rural areas not directly assigned PIM-related responsibilities would also need to be convinced and motivated about the need for them to take a supportive approach to the propagation of PIM.

There is need to ensure speedy universalization of PIM in all irrigation systems and therefore PIM training and orientation needs to receive major attention so as to help in achieving this.

■ Bibliography

GoI, 1996. Report of the Working Group on Participatory Irrigation Management Programme for the IX Five Year Plan. Government of India. Ministry of Water Resources, Command Area Development Wing.

Groenfeldt D., 1998. *Handbook on Participatory Irrigation Management*, EDI, World Bank.

Groenfeldt D. and M. Svendson, 2000. *Case Studies-PIM.* World Bank Institute.

Gupta D. and L.P. Srivastava, 1999. *Status of Participatory Irrigation Management in India: A Documentary Critique.* ISS Manuscript Report 13. Institute of Social Science. New Delhi.

Hooja Rakesh, 1995. *Multidisciplinary Management of Irrigation Schemes and CAD Agencies,*. Occasional Paper No. 2, HCM RIPA, Jaipur.

Hooja Rakesh. 2001. *Multidisciplinary and Participatory Water Management for Command Areas,* Working Paper/Monograph No. 2, India NPIM, New Delhi.

Hooja,, Rakesh, 2002. *A Strategy for Training for PIM in India,* Working Paper/ Monograph No 7, India NPIM, Jaipur.

Hooja,, Rakesh, 2003. "Revisiting Participatory Irrigation Management". *Indian Journal of Public Administration*, Vol. 49, No. 3, July-September 2003

Hooja, Rakesh, 2006. *The Management of Water for Agriculture: Irrigation, Watersheds and Drainage,* Rawat Publications, Jaipur.

Hooja, Rakesh, Anjali Bhatia and Ganesh Pangare, 2001. *Training for Participatory Irrigation Management in India: Present Status and Need: A Policy Aid Document,* Working Paper/Monograph No. 4, India NPIM, New Delhi.

Hooja, Rakesh, Ganesh Pangare and K.V. Raju (eds.), 2002. *Users in Water Management: The Andhra Model and Its Replicability in India,* Rawat Publications, Jaipur.

India NPIM, 2000. *Attitudinal and Behavioural Change for PIM:* A workshop report.

India NPIM, 2001. *Report of the International Training of Trainers Programme on PIM: A Way Forward.*

IRDAS. 1996. *Participatory Irrigation Management: A Handbook for Farmers and Field Functionaries:* See Sithapathi Rao et al., 1999.

Joshi, L.K. and R. Hooja (eds.), 2000. *Participatory Irrigation Management: Paradigm for the 21st Century* (2 volumes). Rawat Publications, New Delhi.

Maloney, C. and K.V. Raju, 1994. *Managing Irrigation Together,* Sage Publications. New Delhi.

Meinzen Dick, R., 1997. "Farmers Participation in Irrigation: 20 Years of Experiences and Lessons for the Future", *Irrigation and Drainage Systems,* Vol.II.

NSPDART, 1998. Proceedings of the Workshop on Attitude and Behaviour Change in Participatory Processes, 24th-28th August, 1998, Ministry of Rural Areas & Employment, GoI, National Society for Promotion of Development Administration. Research and Training, LBSNAA, Mussoorie.

Oza,A., 1997. "Training as a Priority for Participatory Irrigation Management", National Conference on PIM, 20-22 January 1997, New Delhi.

Shah, Anil, 1999. "Capacity Building for Participatory Irrigation Management" Paper presented in the plenary session of National Conference on Participatory Irrigation Management at Hyderabad, 19-23 January, 1999.

Sinha, P.K., 2001, "Education, Training and Other Capacity Building Measures in Water Sector for 21st Century". National Seminar on Land & Water Management Issues for 21st Century, 1-2 March, 2001. Technical Papers, UPWALMI, Lucknow.

Sithapathi Rao, et al., 1999. P*IM: A Handbook for Farmers and Field Functionaries, IRDAS* and WALAMTARI, Hyderabad.

Sivamohan, M.V.K. and C. Scott (eds.), 1994. *India: Irrigation Management Partnerships*, Booklinks, Hyderabad.

SSG, Government of Gujarat, 1999. Report of the Small Study Group Appointed by the Government of Gujarat to Evolve a Plan for Restructuring the Water and Land Management Institute, Gujarat, Small Study Group, Government of Gujarat.

Uphoff, N., Esman, M.J., and A. Krishna, 1998. *Reasons for Success: Learning from Instructive Experiences in Rural Development,* Vistaar Publications, New Delhi.

ANNEXURE I

Indicative lists of topics for training workshops categorized as per various groups that need to be trained (these lists have been taken from Rakesh Hooja, Anjali Bhatia and Ganesh Pangare, *Training for Participatory Irrigation Management in India: Present Status and Needs*, Working Paper/Monograph No. 4, India NPIM India).

1A. Politicians

Training objective: The objective of having orientation programmes for the political leadership is to acquaint them with the developments in the field of PIM in various countries and to equip them with information on the process of irrigation reforms and the political implications thereof.

Knowledge

1. Problems in the present water management
   - Existing irrigation management practices.
   - Analysis of irrigation coverage, efficiency, head and tail end problem, inequity, frequency and reliability of water supply.
   - Financial aspects of the present system.
   - Details on cost effectiveness in O&M – total expenditure and assessment of water fees and recoveries and effect of O&M budgeting.
2. Introduction to PIM
   - Experiences of other counties in PIM.
   - The essential features of PIM.
   - Legal and policy issues in PIM.
   - Expected outcomes from PIM.
3. The implication of PIM
   - Role of Irrigation Department after reform.
   - Role of community organizers/NGOs.
   - Gender issues in PIM.

Skills

1. Steps in PIM
   - Methodology, approach and implementation strategy for PIM

- Planning and monitoring and evaluation tools and processes.

Attitudinal Issues

1. Mental blocks.
2. Apprehensions.
3. Political implications.

IB. Senior Management

Training objectives: For the Irrigation Department, training aims to initiate a process of attitudinal change towards farmers, and for a movement from complete authority over management to joint management of the irrigation systems and to appreciate the role of the farmer association. For other departments (Agriculture, Rural Development, District Administration, etc.), the objective could be an orientation to PIM, and how they could aid the process.

For Irrigation Department

Knowledge

1. Introduction to PIM
   - Experiences of other countries in PIM.
   - The essential features to PIM.
   - Legal and Policy issues in PIM.
2. The implication of PIM
   - Role of Irrigation Department after reform.
   - Role of community organizers/NGOs.
   - Gender issues in PIM,
3. Practical details of the legal back-up
   - Irrigation Act, Canal Rules, Command Area Development Act, etc.,
   - Turnover process, signing MoUs and its implication.

Skills

1. Planning for PIM
   - Legal and policy decision and back-up.
2. Steps in PIM
   - Methodology, approach and implementation strategy for PIM.

- Planning and monitoring and evaluation tools and processes.

3. Management issues
   - Models of WUAs and farmers' organizations at minor, distributary branch level and main canal level.
   - Functions, rights and responsibilities of farmers organizations at each level.
   - Formation and operation of WUAs.
4. Participatory tools and issues
   - PRA and tools for participatory planning and implementation.
   - Climate building for PIM.
   - Conflict resolution.
   - Breaking hierarchy barriers.

Attitudinal Issues (adapted from NSPDART, 1998)

1. Mental block
   - Mistrust in capacity/people.
   - Lack of confidence.
   - Fear of accountability.
2. Apprehensions
   - Income loss.
   - Loss of power
   - Enhanced bargaining position of the farmer.
   - Altered social status.
3. Political implications
   - Deprivation of contract or support.
   - Anxiety arising from uncertainty.
   - Compulsions to spend on O&M.

IC. Middle Management

Training objective: Training to impart knowledge and skills for implementing reform. The training needs to explain the purpose and the inevitability of the reforms, the changes to be expected. It should focus on removing the fears and problems associated with uncertainty in the face of change.

Knowledge

1. Problems in the present water management
   - Existing irrigation management practices.
   - Analysis of irrigation coverage, efficiency, head and tall end problem, insecurity, frequency and reliability of water supply.
   - Financial aspect of the present system.
   - Details on cost effectiveness in O&M – total expenditure and assessment of water fees and recoveries and effect on O&M budgeting.
   - The inevitability of reform.
2. Introduction to PIM
   - Experience of other countries in PIM.
   - The essential features of PIM.
   - Legal and policy issues in PIM.
   - Forms of organizations and WUAs.
3. The implications of PIM
   - Role of Irrigation Department after reform.
   - Role of community organizations and joint management.
   - Role of community organizers/NGOs.
   - Gender issues in PIM.
4. Planning for PIM
   - Legal and policy decisions and back-up.

Skills

1. Aspects of joint management
   - Joint inspection, fixing water quotas, evolving volumetric rates, rehabilitation of irrigation network as per users vies, hydraulic tests of the system before handing over, etc.
2. Practical details of the legal back-up
   - Irrigation Act, Canal Rules, Command Area Development Act, etc.
   - Turnover process, signing MoUs and its implication.
3. Participatory tools and issues
   - PRA and tools for participatory planning and implementation.

- Climate building for PIM.
- Conflict resolution.
- Breaking hierarchy barriers.

Attitudinal Issues

1. Mental block
   - Mistrust in capacity/people.
   - Lack of confidence.
   - Fear of accountability.
2. Apprehensions
   - Income loss.
   - Loss of power.
   - Enhanced bargaining position of the farmer.
   - Altered social status,
3. Political implications
   - Deprivation of contract or support.
   - Anxiety arising from uncertainty.

ID. Section Officers/Field Staff

1. Introduction to PIM
   - Experiences of other countries M PIM.
   - The essential features of PIM.
   - Legal and policy issues in PIM.
2. The implication of PIM
   - Role of Irrigation Department after reform with emphasis on the role of the field staff.
   - Role of community organizers/NGOs.
   - Gender issues in PIM.
3. Forms of organization
   - Responsibilities and duties at various levels.

Skills

1. Water User Association
   - Formation of WUA – its, rights and responsibilities.
   - Registration of WUAs, conducting meetings, joint inspections.
   - Record keeping and reporting.
   - Conflict resolution.

2. Technical issues in joint management
   - Hydraulic tests.
   - System operation and rotational water supplies.
   - Water deliveries to minors, water allocation principles, water-measuring devices, control structures-locations, nature, conditions, accuracy and maintenance policy.
3. Legal issues in PIM
   - Canal Rules, Command Area Development Act, etc.
   - Turnover process, signing MOUs and its implication.
4. Monitoring and evaluation for supervisory capability for activities carried out by the WUAs
   - Distribution of water.
   - Rules for water distribution.
   - Scheduling of irrigation.
   - Water accounting.
   - Annual financial budget and water budgets.
   - Book-keeping and auditing the expenditure.
   - Maintenance of the system.

Attitudinal Issues

1. Mental block
   - Mistrust in capacity/people.
   - Lack of confidence.
   - Fear of accountability.
   - Participatory techniques.
   - Communication with farmers – how to approach farmers, how to talk with them, motivation, etc.
2. Apprehensions
   - Income loss.
   - Loss of power.
   - Enhanced bargaining position of the farmer.
   - Altered social status.
3. Political implications
   - Deprivation of contract or support.
   - Anxiety arising from uncertainty.

IE. Farmers

Knowledge

1. Water User Associations
   - Formation of WUAs – their rights and responsibilities and functions.
   - Functions and responsibilities of the Irrigation Department.
   - Functions of office bearers, executive committee, general body of the organizations at minor, distributary, branch level and main canal level.
   - Registration of WUAs
   - Procedures for conducting meetings and decision making of WUAs
   - Record-keeping and reporting.
   - Conflict resolution.
   - Fixing water fees for different crops to meet management costs of WUAs.
2. Legal issues in PIM
   - Irrigation Act and Canal Rules, Command Area Development Act, etc.
   - Turnover processing MoUs and its implication.
   - By-laws of WUAs MoU between Irrigation Department and WUAs

Skills

1. Technical issues in joint management
   - Existing practices of O&M by Irrigation Department.
   - Joint inspection for rehabilitation.
   - Hydraulic tests.
   - Preparation of annual water budget.
   - Water deliveries to minors, water allocation principles, water measuring devices, control structures-locations, nature, conditions, accuracy and maintenance policy.
   - Environmental aspects.
   - System operation and rotational water supplies.
   - System maintenance: Maintenance of repairs of canals, field channel and drain maintenance.

2. Activities carried out by the WUAs
   - Distribution of water.
   - Rules for water distribution.
   - Scheduling of irrigation.
   - Water accounting and water budgets.
3. Accounting and financial management Annual financial budget.
   - Book-keeping, balance-sheet, profit and loss accounts.
   - Auditing the expenditure.
   - Maintenance of the system.

Attitudinal Issues

- Leadership.
- Self-organization-group functioning.

IF. NGOs

1. Organizing training programmes.
2. Training needs assessment.
3. Participatory tools: Participatory development of training material.
4. Overview of PIM in Indian states and other countries.

Chapter 10

Community Managed Drinking Water Systems in the Himalayas

An NGO Perspective

Kalyan Paul

In the central and western Himalayas, over 250 hamlets have been managing drinking water supply systems, through the adoption of an appropriate technology application in the form of infiltration wells, which act as a protected intake structure, fitted with handpumps or submersible power lift pumps. This note documents the remarkable manner in which communities have been enabled to plan, implement, operate and maintain drinking water and environmental sanitation systems.

Drying up of local water sources has created severe hardships for a majority of people residing in the central and western Himalayas. Traditional methods of intercepting shallow sub-surface water capillaries into *naulas* and *bauris* are mostly inadequate, as the denudation of vegetal cover in the catchment areas has resulted in decreasing the annual recharge of the primary water base. Besides this, the population in each hamlet has also increased to extents that dwindling water resources cannot provide enough for everyone's needs.

In the past three decades, chronic water shortages have resulted in diverting large sums of public funds for carrying piped water over several miles from distant secondary and even tertiary water sources to the problem villages. Such an approach does not meet the sustainability criteria on four major counts:

- The cost-benefit ratio is skewed due to a per capita investment of over Rs 5,000.
- It diverts the attention of the local people regarding the critical need to reforest and protect the village catchment areas.
- Operation and maintenance responsibilities are controlled by external technocratic departments with huge annual recurring costs.
- And, in the long run, with increasing biotic pressure in denuded catchment areas wherein secondary water sources are located, the problem of water shortages seem far from being resolved.

In view of this, *Grassroots* (an NGO) has found it relevant to introduce an appropriate technology application, conceptualized and developed by Dr Tim Rees, through the construction of infiltration wells and installation of handpumps, power lift pumps and gravity flow water supply systems.

This appropriate technology application has several advantages:

- It is based on local primary water resources and may assist villagers in realizing the benefits of protecting the immediate catchment areas.
- Since deeper seepage lines are intercepted (from the traditional *naula bauri* of 5-6 feet to 30-40 feet in the case of infiltration wells), it is possible to harness enhanced quantities of drinking water.
- Chances of external contamination is avoided as the infiltration well is located several feet underground and covered.
- Per capita cost is less than Rs 500 as compared to Rs 5,000 for conventional systems and almost negligible operation and maintenance costs, which is in any case, viewed as a local responsibility, albeit with appropriate training inputs.

- Reduces drudgery of women and children regarding collection of water from distant sources.
- Leads to improved personal hygiene and sanitation due to availability of enhanced quantities of water.

In the last twelve years, about 35,000 people of 250 communities in far-flung micro watersheds in the hill states of Uttaranchal and Himachal Pradesh have gained access to enhanced quantities of safe drinking water through adoption of this appropriate technology application. All these communities not only contributed ten per cent of the costs upfront but also agreed to bear the responsibility of operation and maintenance.

Field Examples: Community Management of Appropriate Technology

At Panergaon Village

This community was part of a multi-village water supply scheme, based on a distant secondary water source located in a reserved forest. Ironically, the state of the 'reserved' forest deteriorated, which led to the drying up of the water source to the extent that the scheme was defined as non-functional. All that was left of the huge public investment was miles of dry pipelines. And, typically, the village was re-defined in the problem village category and thereby eligible for further public investments! The story of Pangergaon village is typical of large swathes of land and people in the Himalayas.

Collecting even small quantities of water was a regular chore for the people in the village, who had to trudge up and down the hillside for over six hours daily. The quality of such water was also dubious and led to chronic ailments.

During the onset of the first phase of the drinking water sector reform programme, the villagers decided to take matters in their own hands. Within a year, the community had prepared an ambitious planning document regarding water and environmental sanitation and decided to entrust the responsibility of implementation, operation and maintenance to a democratically selected Village Water & Sanitation Committee (VWSC).

Over the next year or so, the VWSC sought necessary assistance from Grassroots and managed to accomplish the following

- Construct an Infiltration Well which harvested over 20 litres of water per minute or close to 30,000 litres per day from the primary water resource base within the village jurisdiction.
- Install submersible pumpsets to lift the water up to the top of the village by 90 metres, into pressed steel pre-fabricated storage tanks lined with fibre glass.
- Provide enhanced quantities of safe potable water to the entire community through a network of gravity-fed pipelines, both as public standposts and private connections.
- Reduce the drudgery of women regarding collection of water and thereby enable them to spend time more meaningfully.
- Motivate every household to construct twin pit waterseal *sauchalayas* in order to prevent contamination of water through open defecation and thereby improve the level of personal hygiene significantly.
- Ensure that each household contributed up to 10 per cent of the capital cost.
- Set in place a sustainable management system regarding operation and maintenance of the water supply system through regular collection of water cess.
- And, finally, initiate a process of protection and conservation of the critical catchment area of the village.

At Chamni Village

This village also had been covered by the state government through a gravity flow water supply scheme; which, unfortunately, had been functional only for a short period of less than one year.

With some help from *Grassroots*, the community constructed an Infiltration Well right next to the defunct intake structure (boulder filled gallery) and harvested sufficient water for two overhead storage tanks. Thereafter, through a network of underground pipes, water was distributed to each and every household through a system of public standposts.

Alongside, in order to further enhance the availability of water, the community decided to construct:

- four more infiltration wells fitted with handpumps at various locations,

- over forty individual household level rainwater harvesting structures, and
- cent per cent coverage of households with twinpit waterseal *sauchalayas.*

The people of Chamni have realized the benefits of cooperation to the extent that fresh vegetal matter in the form of mature saplings of local shrub and tree species have been planted-out on the denuded catchment areas surrounding the village.

At Dhankoli Village

After the successful demonstration of an Infiltration Well as a protected intake structure for gravity-flow water supply system at Chamni, another community decided to adopt similar technology for resolving their water problems. It is significant to note that two infiltration wells could be located within the village jurisdiction, in comparison to the earlier non-functional scheme which had attempted to carry piped water from a distance of over 10 kms.

At Malona, Nausar and Rawalsera Villages

Based on the success of the water management systems in Panergaon, three other communities decided to opt for similar technology application – infiltration wells fitted with submersible pumps.

Besides these examples, *Grassroots* has been involved with enabling communities in over 200 other hamlets to plan, implement, operate, maintain and share the cost of such community managed drinking water systems.

Concluding Remarks

It is significant to note that all these communities are not only proud owners of water supply systems but are also more than able to operate and maintain the systems quite independently for the past several years.

The tragedy is the government is still far from being convinced that such possibilities actually exist and need to be encouraged. The main factor for such a perspective is probably due to the technocratic institutions of the state governments which have taken it upon themselves to design and implement water supply schemes with absolutely no participation of' the community.

In general, it is apparent that communities are more than able to manage their lives, provided they are provided with the opportunity and freedom to do so. Of course, villagers need to be drawn into innovative dialogues regarding ideas and change, in order to enable them to decide upon the ideal choice of technology.

Discussions regarding 'choice of technology' need to be facilitated through farmers induction programmes, whereby people from problem-villages ought to be encouraged to spend day or so on a study tour of another village where the community has already taken the first steps towards freedom and development of community managed drinking water systems.

Finally, it may be in order to mention that the *Grassroots* has been able to provide such support systems to communities largely due to the creation of an *Artisans Guild* which comprises of almost forty local youth trained as barefoot engineers. The barefoot engineers play a key role in enabling communities to adopt appropriate technology options not only in the drinking water sector but also in other cross cutting sectors like environmental sanitation, renewable energy, rainwater harvesting and catchment area treatment.

Chapter 11

Watershed Approach

A Viable Proposition for Water Management in Rainfed Areas

C. Sekar and S. Senthilnathan

THE importance of rainfed crop production is partly due to the dominance of rainfed agriculture in developed countries. More than 80 per cent of cereal area in developed countries is rainfed, much of which is highly productive maize and wheat land such as that in the mid-western US and parts of Europe. The average rainfed cereal yield in developed countries was 3.2 metric tonnes per hectare, virtually as high as irrigated cereal yields in developing countries. Rainfed cereal yields in developed countries are projected to grow to 3.9 metric tonnes per hectare by 2021-25. Irrigation is relatively more important in cereal production in developing countries. Rainfed yields in developing countries are projected to increase from 1.5 metric tonnes per hectare to 2.1 metric tonnes per hectare by 2021-25, and rainfed area in developing countries will account for 43 per cent of total cereal area, and rainfed areas will account for 40 per cent of growth in cereal production. Globally, 69 per cent of cereal area is under rainfed including 40 per cent of rice, 66 per cent of wheat, 82 per cent of maize, 86 per cent of other grains and 85 per cent of soybeans. The global rainfed harvested area of rice, wheat, maize, other cereals, soybeans, potatoes, sweet potatoes, and cassava and other roots is 560 million hectares in 1995 with cereals representing 85 per cent of the total. Worldwide, rainfed cereal yield is about 2.2 metric tonnes per hectare, which is about 65 per cent of the irrigated yield. Rainfed cereal production accounts for 58 per cent of global cereal production. The global rice harvested area is 146 million hectares of which approximately 87 million hectares are irrigated, and 59 million hectares are rainfed. There is very little rainfed rice planted in developed countries, while rainfed rice occupies approximately 42 per cent or 59 million hectares of the total rice area in developing countries. Developing countries are also responsible for almost the entire world production, with 97 per cent of the total world rice yield coming from developing countries. Rainfed rice-yield in developing countries is 1.4 tonnes per hectare or about 44 per cent of the total irrigated rice yield in developing countries and this amounts to 24 per cent of the developing country total, and 23 per cent of world production (IFPRI Report, February 2002).

For instance, about 222 million hectares of wheat was harvested worldwide in 1995, about 66 per cent of which was rainfed, and the remaining 34 per cent irrigated. It is interesting to know that about 83 per cent of the area under wheat in developed countries was rainfed, while in developing countries less than half of the total wheat area planted was rainfed. Rainfed wheat yields in developed and developing countries are approximately 2.5 tonnes per hectare and 1.2 tonnes per hectare respectively while the irrigated yields are slightly higher at 2.9 tonnes per hectare and 1.7 tonnes per hectare. Rainfed wheat production contributes 33 per cent of the total yield in developing countries, 81 per cent in developed countries, and 52 per cent worldwide. Maize is grown under rainfed conditions more often than rice and wheat. Of the total 138 million hectares of maize in the world, 82 per cent is rainfed, while 18 per cent is irrigated. Developing countries occupy over 60 per cent of the total maize area worldwide. The average rainfed maize yield in developed countries is 3.4 tonnes per hectare, while that of developing countries lags behind at 1.8 tonnes per hectare. Irrigated yields are higher at 4.2 tonnes per hectare in developed countries and 2.9 tonnes per hectare in developing countries. Rainfed maize production contributes 66 per cent of the total yield in developing countries, 81 per cent in developed countries, and 74 per cent globally. It is a matter of fact that just about 62 million hectares of soybeans are harvested worldwide, of which 53 million hectares are rainfed. Developed countries plant 91 per cent of the total soybean area using rainfed agriculture, while 80 per cent of the area in developing countries is rainfed. Unlike cereal crops, rainfed and irrigated soybean yields are similar. In developed countries, the irrigated soybean yield is 2.7 tonnes per hectare, higher than the rainfed yield of 2.2 tonnes per hectare. In developing countries, the irrigated yield is only marginally higher than the rainfed yield, with both at approximately 1.8 tonnes per hectare (IFPRI Report, February 2002). The foregoing discussions clearly explicate the importance of rainfed areas in the food production front and its contribution in the global food basket.

Rainfed Farming vs. Irrigation Investments

A number of alternative scenarios show that more rapid growth in rainfed yield and production could compensate for reduced investments in irrigation or reduced ground water pumping to eliminate groundwater overdraft, but achieving the required improvements in rainfed production would be a significant challenge. The estimated rate of groundwater extraction in the 1990s, exceeding the replenishment rate has been calculated at 104 billion m3yr-1 compared to 30 billion m3yr-1 in China and 10 billion m3yr-1 in northern Africa. Thus, for example, a scenario that eliminates groundwater mining throughout the world would result in a decline in irrigated cereal production of 20.1 million metric tonnes in China, 18.4 million metric tonnes in India, 18 million metric tonnes in WANA, 1.6 million metric tonnes in developed countries, and 53 million metric tonnes in developing countries as a whole in 2021-25 relative to the baseline. These reductions can be offset by an increase in rainfed area and yield, but the required increase in yields would be very large. Compared to the baseline, average rainfed cereal yield would need to increase by 13 per cent or 0.6 metric tonnes per hectare in China, 20 per cent or 0.30 metric tonnes per hectare in India, and 0.3 metric tonnes per hectare in WANA and rainfed cereal area will increase by 0.6 million hectares in China, 0.8 million hectares in India, and 0.10 million hectares in WANA (IFPRI Report, February 2002). However, irrigation accounts for about 72 per cent of global and 90 per cent of developing country water withdrawals, and water availability for irrigation may have to be reduced in many regions in favour of rapidly increasing nonagricultural water uses in industry and households, as well as for environmental purposes. The concern over increasing water scarcity for irrigation, the role of water management and investments for irrigated agriculture and food security has received substantial attention in recent years (Hofwegen and Svendsen, 2000).

Status of Rainfed and Irrigated Agriculture in 2025

A comparison of the average projected rainfed and irrigated cereal area, yield, and production during 2021-25 with the same items for 1995 shows changes in these items, a reflection of the changing role of rainfed agriculture during 1995-2025 under the baseline scenario. Worldwide, rainfed cereal area in 2021-25 is projected to be 512 million hectares, an 8 per cent increase over the area in 1995. Increases in irrigated area, yield and production by 2021-25 are expected to be slightly higher than rainfed increases with an 11 per cent increase in irrigated area expected over 1995 levels, a 38 per cent increase in yield projected, and a 53 increase in production expected over 1995 values. Rainfed cereal area will still account for the majority of total harvested area and total cereal production in 2021-25, although irrigated area and production comprise slightly more of the totals than in 1995. The developing world maintains very similar fractions of rainfed area and production to the values in 1995, and experiences a yield growth of 0.57 tonne per hectare. The developed world will experience slightly larger increases in the fraction of rainfed area and rainfed production, as well as a yield growth of 0.72 tonne per hectare over 1995 levels (IFPRI, 2002).

Rainfed Farming and its Importance in India

Rainfed agriculture in India is complex, diverse and risk-prone and is characterized by low levels of productivity and low input usage. Vagaries of the monsoon result in wide variation and instability in yields. Majority of the rural poor live in rainfed regions. The rainfed agriculture plays an important role in the Indian economy. India shares about 16 per cent of the global population but it has only 4 per cent of the world's total water resource (Government of India, Planning Commission, 2001). In India, 68 per cent of total net sown area (about 136.8 million hectares) is under rainfed lands, spread over 177 districts. Rainfed crops account for 48 per cent of food crops, 75 per cent of oilseeds, 90 per cent of pulses, 70 per cent of cotton and 68 per cent of non-food crops. Nearly 50 per cent of the total rural workforce and 60 per cent of livestock in the country are concentrated in the dryland regions. Currently, over 10 per cent of blocks, or areas classified by the Central Ground Water Board of India have been identified as being overexploited and blocks where exploitation is beyond the critical level have been increasing at a rate of 5.5 per cent each year (World Bank, 1999). In Tamil Nadu, out of 7 million hectares of cultivable area, around 3.1 million hectares (44%) are under rainfed agriculture. But uncertainty in production due to fluctuations in total rainfall and changes in its distribution, decrease in relative productivity in rainfed lands, etc. affects the livelihood security of many poor and marginal farmers.

Rainfed Agriculture: Challenges

Rainfed and irrigated agriculture coexists practically in every village in India. Hence, public investment programme cannot be targeted so precisely. For practical purposes, they need to be location specific planned and implemented on a larger scale, such as at the village, *taluk*, district or state level. About 70 per cent of arable lands in India is rainfed and depends on monsoon rains and 45 per cent of these drylands still continue to depend on natural precipitation even after realizing the full irrigation potential. It is estimated that the agricultural growth in these rainfed areas had been relatively low. The main source of instability in the overall agricultural production in India is due to unevenness in the production scenario of dryland agriculture. This unexciting picture of dryland agriculture is a major cause of anxiety because it is the natural habitat of most of the pulses and oilseeds, which are in short supply but are crucial for the Indian economy. The challenge before Indian agriculture is to transform rainfed farming into more sustainable and productive systems and to better support the population dependent upon it. The green revolution, considered to be the cornerstone of India's agricultural growth by-passed the rainfed areas, remaining confined primarily to irrigated tracts. With productivity levels of staple crops in the irrigated areas plateau ring off, the ability of the traditional bread-basket areas to continue to fill the grain basket is in question. If food production is to be doubled in the next decade, all areas where rainfed farming is predominant will need to contribute substantially.

Degradation of Natural Resources

Rainfed areas in India has considerable production potentials, which have not yet been fully exploited due to restricted availability of water and low soil fertility. Hence, productivity in rainfed areas remained not only low but also greatly fluctuating from year to year. Non-receipt of sufficient amount of rainfall at the time of requirement, ineffective distribution of rainfall, flooding for a short span of time etc., have led to poor infiltration of runoff water, which is essential for groundwater recharge. The basic problem of rainfed areas is degradation of natural resource base like soil through poor management leading to low productivity, low income and low surplus. This, in turn, leads to overexploitation of the existing natural resource base like groundwater. Among the natural resources that support life in all forms, land is the most crucial resource as it provides the base for storing surface and underground water and also for vegetation, which mismanaged or neglected their mother land, had either to migrate *en mass* or else they gradually decline and their civilizations are extinct now (Lal, 2001).

Soil and Water Conservation in Rainfed Areas: Key Issues

Land and water are the two important and most fundamental resources for the survival of living beings. It is highly disturbing to find that because of growing human population all these resources are fast depleting. In addition to this, it threatens not only human prosperity but also the very survival of the human race due to unresponsive use of the natural resources. The situation demands all-out efforts towards finding practical solutions to the above problems. Therefore, in order to obtain sustainable agricultural production as well as to improve soil productivity and ecosystem of vast rainfed agriculture, the efforts to conserve as much rainwater and to prevent soil erosion are considered critical. This can be implemented effectively and economically by adopting various soil and moisture conservation activities on cost effective integrated approach. There is no denying the fact that green revolution of the 1960s has centred mostly on irrigated areas, and yield gaps between irrigated and rainfed areas were quite substantial. But evidences indicate that even farmers in rainfed areas have also adopted HYV technologies, particularly where there is assured rainfall and the occurrence of floods and droughts are minimal. Nevertheless, the yield response to technology adoption in rainfed areas is much less, as compared to irrigated areas. Also, the yield variability is relatively higher. Due to low and uncertain yield response to new technology as well as low capital absorption capacity, rainfed areas have lagged behind in terms of both technology adoption and productivity growth. As a result, the rainfed areas are associated with higher incidence of poverty. However, the fact is that there is a limit beyond which irrigation potential cannot be increased and that irrigated areas have already reached a level in the adoption of available new technology, do compel us to look for alternative sources of agricultural growth, particularly in rainfed areas. In fact, India will have to increasingly harness the potentials of rainfed agriculture for ensuring the food and nutritional security of its growing population. Although the present food supply situation looks comfortable, the growing demand for foodgrains to meet the demand of increasing population and rising income of the poor may compel us to look for alternative sources and areas of food production.

Conservation of Water: Need of the Hour

Majority of the Indian villages have traditionally relied on local rainfall for meeting the water requirements of agriculture and domestic activities. An array of techniques, adapted to local conditions, have been evolved and adopted over centuries for harnessing rain water and conserve it in small surface storages and underground. These are still in active use but to some extent they have been overshadowed by large reservoirs and energized pumping of groundwater which is made possible by technological advancements. In India, the quantum of water harnessed by these means has expanded phenomenally and now far exceeds the quantity obtained by traditional devices. However, it is becoming apparent that (i) the demand for water for all uses has grown largely and continues to grow at an unprecedented rate, (ii) while supplies have increased, all parts of the country have not benefited uniformly from it, (iii) nearly two-thirds of the country's cultivated land still depends entirely on rainfall, (iv) a large proportion of the population does not have adequate good quality water, (v) the scope for augmenting supplies from surface sources is limited and diminishing even as there are clear signs that groundwater resources are being already over exploited and subjugated, and (vi) competition between different uses and among users for the limited supplies of water has been greater than before in many areas. In this context, conservation of soil and prudent use of water from reservoirs and underground are imperative and warranted.

Watershed Approaches

The conservation of land and water resources in rainfed areas has been neglected for too long. In order to conserve the precious natural resources, viz., soil, water etc. and also to sustain the existing resource endowments for future entitlement, the Government of India has formulated "Watershed Approaches". The basic objectives of the watershed programme among other things are location specificity, resource conservation, environmental and ecological concerns. The available information on watershed shows that the watershed approaches have already paid dividends and the intended objectives of conserving soil, moisture, arresting soil erosion, improving crop productivity, enhancing the overall socio-economics of the farmers, attaining environmental and ecological sustainability etc. are fulfilled. Watershed in short is a self-defended area, which does not allow any water from outside the catchments to enter into it and allow its excess water to discharge to a common point in a stream. The basic concept of watershed management is to conserve all the basic natural resources of the watershed and to plan for their optimal utilization, the approach and strategy are primarily based on the twin concepts of (1) integrated watershed management, and (2) sustainable farming systems. A watershed consists of three physical sectors: (i) arable or cultivated lands, (ii) non-arable lands-village pastures and grazing ground, culturable wastelands and barren and uncultivable lands, and (iii) network of natural drainage lines.

Components of Watershed Development

Watershed development refers to the conservation, regeneration and the judicious use of all the resources, natural (land, water, plants and animals) and human, within a particular watershed. Watershed management tries to bring about the best possible balance in the environment between natural resources on the one side, and human and other living beings on the other. All these components are interdependent and interactive. The watershed approach enables a holistic development of agricultural and other allied activities. The key attributes of watershed management are conservation of rainwater and optimization of soil and water resources in a sustainable and cost-effective mode. It aims to optimize moisture retention and reduce soil erosion, thus maximizing productivity and minimizing land degradation. Improved moisture management increases the productivity and hence conservation and productivity enhancing measures are complementary. In 1999-2000, for the first time, a Watershed Development Fund (WDF) was established at NABARD on the lines of the Rural Infrastructure Development Fund (RIDF) to enable state governments to access credit for treatment of larger areas under watershed development. Impact evaluation studies both on the ground and through remote sensing techniques have shown that watershed based interventions have led to increases in ground water recharge, increase in water levels of wells and water bodies, enhancement of cropping intensity, changes in cropping pattern, higher yields of crops and reduction in soil losses.

The restructured, second generation, watershed projects have kept people's participation centre stage. It has now become mandatory for watershed development to be planned, implemented, monitored and maintained by the watershed community themselves. Three decades after the advent of the green revolution, watershed development in rainfed areas is gaining the same momentum and appears to be on its way to becoming a people's movement. The objectives of watershed development are such as minimizing ecological degradation, reducing regional disparity and also opening up of greater opportunities for employment of the rural poor in the rainfed areas. The watershed framework thus provides the necessary inter-ecological linkages and is a logical unit for the integration of the sustainable use of land and water. It helps augment soil and groundwater resources. It also enjoins the biophysical, social and economic inputs, which if optimally managed, lead to diversified high agricultural production, control of environmental degradation and provide a mechanism for the recharge of groundwater aquifers. Watersheds thus ensure water adequacy to crops and reduce the risks of yield loss due to water related stresses.

Watershed Development Programmes in India

Watershed development approach was first adopted in a significantly large scale in 1974 in the catchments of a few selected river valley projects. The major watershed development projects currently underway in India include 47 Model Watershed Projects of the ICAR (Indian Council of Agricultural Research), a centrally sponsored scheme of Integrated Watershed Management in the Catchments of Flood-Prone Rivers. World Bank-funded watershed development projects, in the states of Andhra Pradesh, Karnataka, Madhya Pradesh, Maharashtra and Tamil Nadu, the District Watershed Development Programmes of Karnataka, and again a centrally-sponsored scheme of National Watershed Development Project for Rainfed Area. The National Watershed Development Project for Rainfed Areas (NWDPRA) was launched during the Eighth Plan (1992-1997) in 25 states and two union territories and continued to be implemented during the Ninth and Tenth Plans. The broad development objectives of the projects are: (i) enhancement of agricultural productivity in a sustainable manner, (ii) restoration of ecological balance in the degraded and fragile rainfed ecosystems by greening these areas through appropriate mix of trees, shrubs and grasses, (iii) reduction in regional disparity between irrigated and rainfed areas, and (iv) creation of sustained employment opportunities for the rural poor. Impact evaluation studies conducted both on the ground and through remote sensing techniques have shown that watershed based interventions have led to increases in groundwater recharge, increase in number of wells and water bodies, enhancement of cropping intensity, changes in cropping pattern, higher yields of crops and reduction in soil losses.

The National Watershed Development Project for Rainfed Areas (NWDPRA), which is a centrally-sponsored scheme, is under implementation in Tamil Nadu since 1990-91. The twin objectives of the scheme are sustainable production of biomass and restoration of ecological balance in the watershed areas. The scheme is being implemented in 84 watersheds spread over 19 districts, viz., Cuddalore, Villupuram, Vellore, Salem, Thiruvannamalai, Namakkal, Dharmapuri, Coimbatore, Erode, Tiruchirappalli, Karur, Perambalur, Pudukottai, Madurai, Theni, Dindigul, Sivagangai, Virudhunagar and Tirunelveli (refer Table 11.1 for details).

Table 11.1

Fund Outlay for 1999-2000 from NWDPRA

Name of the District	*Block/Watershed*	*Project Area (ha)*	*Project Cost (Rs in lakhs)*
Salem	Alathur	5000	200.00
Namakka	Sentha mangalam	5105	204.20
Theni	Vaigai, Manjalar, & Suruliyar	5151	206.04
Ramanathapuram	Nainarkoil	5060	202.40
Virudhunagar	Vembakottai	5300	212.00
Karur	Kadavur	11606	464.24
Cuddalore	Gadilam	5580	223.80
Total		42802	1719.68

Drought-Prone Areas Programme (DPAP) is yet another important programme implemented with the prime objective of promoting the overall economic development of the watershed community by putting the natural resources like land, water, etc., to the optimum use so as to mitigate the adverse effects of drought and prevent further ecological degradation, besides employment generation through non-farm activities. The Government of India has formulated new guidelines during 1995-96 shifting the focus to watershed development approach with people's participation. In Tamil Nadu state, the programme is implemented in 80 blocks of 16 districts, viz., Dharmapuri, Thoothukudi, Sivagangai, Ramanathapuram, Virudhunagar, Pudukottai, Tirunelveli, Salem Namakkal, Coimbatore, Tiruvannamalai, Dindigul, Vellore, Tiruchirappalli, Perambalur and Karur. The expenditure towards the programme was shared equally between the Central and State government on 50:50 basis upto 31.3.1999. However, from 1 April 1999, the expenditure is shared at the ratio of 75:25 for new watershed projects sanctioned by the Government of India. The allocation of funds for watershed activities during different phases is given in Table 11.2.

Table 11.2

Fund Outlay and Expenditure for DPAP

Watershed Project Batch No. & Period of implementation	*Project Outlay (Rs in lakhs)*
Watershed Projects (1 batch) 1995-96 to 1999-2000	5940
Additional Watershed Projects (IV batch) 1997-98 to 2000-2001	2060
New Watershed Projects 1999-2000 to 2002-2003	5980

Impact of Watershed Activities in a Rainfed Area: A Field-Level Experience

A micro-level study was conducted in 2003-04 to analyze the impact of watershed programmes on the socio economic characteristics of the user groups (farmers) in Pongalur block of Tiruppur taluk in Coimbatore district, where the watershed activities are carried out by the development departments in association with the farmers (for details, refer Senthilnathan and Sekar, 2004). The IWDP was implemented in the study area on a large scale. The various watershed developmental activities undertaken in the study area are percolation pond, minor check dam, major check dam, land levelling, farm ponds, contour bunding, summer ploughing, agro forestry etc. The overall objective of the study was to assess the cost economics of different resource conservation technologies and also to study the comparative performance of agricultural and other socioeconomic parameters between WTA (watershed treated area) and NTA (non-treated area) through indicators like cropping pattern, cropping intensity, income inequality etc.

General Characteristics of the Study Area

The study was conducted purposively in Pongalur block of Coimbatore district in Tamil Nadu because of the following two reasons: (i) This block was the site of implementation of watershed under the Integrated Watershed Development Programme (IWDP). (ii) The study area was also the location of many watershed treatment activities, which include among other things the soil and moisture conservation activities like summer ploughing, land levelling and contour bunding. This block has eighteen watersheds spread over six panchayats, viz., Elavanthi, Kattur Madappur, Pongalur, Vadamalipalayam and V.Kallipalayam. The study mainly focused its attention on the comparative analysis of costs and returns between WTA and NTA. Hence, the present study required samples of the beneficiaries from WTA and non-beneficiaries from NTAs. Thus, to enable a comparative study, the total number of respondents was fixed at one hundred and twenty. Among the six panchayats in the block, Kattur panchayat was randomly selected for the study. From the above panchayat, two villages for WTA and two villages for NTA were selected randomly. From the villages, two broad categories of 60 beneficiaries from WTA and 60 non-beneficiaries from NTA were selected randomly for the study.

Cost and Returns and Productivity of Crops

It was understood that the cropping intensity of the sample farms in WTA was higher (114.31%) than the sample farms in NTA (102.90%). Enhanced cropping intensity in WTA could be attributed by the meticulous implementation and adoption of watershed technologies. The comparative costs and returns of sample farms between WTA and NTA were analyzed by calculating the cost and returns of some of the dominant crops discretely and the results are presented in Table 11.3. The percentage difference in costs and returns was also computed individually to understand the efficacy of the conservation compliance. It could be known from the table that coconut, maize, sorghum, chillies and onion had shown a positive additional net return of 27.35, 31.68, 13.06, 24.54 and 28.06 per cent, respectively. Thus, it is clear that the net returns of major crops in the study area had been higher in the sample farms in WTA than the sample farms in NTA.

Table 11.3

Comparative Study of Cost and Returns of Major Crops in WTA and NTA

(in Rs)

Particulars	*Coconut*		*Maize*		*Chillies*		*Sorghum*		*Onion*	
	WTA	*NTA*	*WTA*	*NTA*	*WTA*	*NTA*	*WTA*	*NTA*	*WTA*	*NTA*
Preparatory Cultivation	-	-	537	513	472	422	550	500	554	500
Labour	2854	2120	2695	2428	2844	2520	800	730	3110	2565
Seed	-	-	693	450	211	195	530	470	1359	1124
Manures	2065	1875	1200	900	1005	982	-	-	381	250
Fertilizers	1489	1290	922	679	1025	910	-	-	1059	785
Plant Protection	1472	854	250	200	344	254	-	-	404	352
Total Cost	8291	6139	6046	5169	6245	5286	1980	1700	6869	5577
Total Return	61212	47692	20967	16500	21456	17500	5010	4380	22772	17996
Net Return	52921	41553	14920	11330	15211	12213	3030	2680	15903	12418
Percentage Difference (Net Return)		27.35		31.68		24.54		13.06		28.06

Productivity and Yield Difference

The productivity of major crops and their yield difference were analyzed to study the differences in yield among the sample farms in the WTA and NTA. The results are shown in Table 11.4. It could be observed from the Table 11.4 that the productivity of maize showed the highest increase in productivity (16.08%) followed by coconut (15.39%).

Table 11.4

Productivity of Major Crops and their Yield Difference

(in kg/ha)

Crops	*W TA*	*N TA*	*Percentage of Yield Difference*
Coconut (in numbers)	14170	12280	15.39
Maize	4869	4194	16.08
Green Chillies	10640	9350	13.00
Onion	31710	28801	9.20
Sorghum	747	694	7.64

It is obvious that the crops like chillies, onion and sorghum also indicated increase in productivity to an extent of 13 per cent, 9.20 per cent and 7.64 per cent, respectively. Thus, it could be concluded that watershed development programme had a significant impact on the productivity of the crops.

Water Quality

The water quality of the sample farms in the watershed was assessed through collection of opinions from the sample respondents. The respondents' opinion for different quality attributes of water are tabulated for easy understanding. From the Table 11.5, it could be interpreted that about 37 per cent of farmers in WTA perceived that the water quality had been good, while only 21.67 per cent of farmers in NTA observed that the quality of water was good, which shows the indirect benefit received by the adopters of

Table 11.5

Opinion of the Sample Farmers on Water Quality

Water Quality	*WTA*		*N TA*	
	Number	*Percentage*	*Number*	*Percentage*
Good	22	36.67	13	21.67
Medium	28	46.66	26	43.33
Poor	10	16.67	21	35.00
Total	60	100.00	60	100.00

soil and water conservation technologies in WTA. About 47 per cent of the farmers in WTA perceived that the water quality was medium, while only 43.33 per cent of the farmers in NTA reported that the water quality was medium. Thus, it could be concluded that the overall quality of water in WTA had been superior to NTA.

Land Quality

The land quality of the sample farms in the watershed was assessed through collection of opinions from the sample farmers. The respondents' judgment for different quality attributes of land are reported in Table 11.6.

Table 11.6

Opinion of the Sample Farmers on Land Value

Land Value	*WTA*		*NTA*	
	Number	*Percentage*	*Number*	*Percentage*
Good	23	38.33	16	26.67
Medium	25	41.67	21	35.00
Poor	12	20.00	23	38.33
Total	60	100.00	60	100.00

From the Table 11.6, it could be understood that about 58 per cent of farmers in WTA perceived that the land value was good, while only 26.67 per cent of farmers in NTA reported that the quality of land was good, which showed the circumlocutory benefit received by the adopters of soil and water conservation technologies in WTA. About 42 per cent of the farmers in WTA perceived that the land quality was medium while only 35 per cent of the farmers in NTA revealed that the land value was medium. Thus it could be reported that the overall quality of land in WTA was superior to NTA.

Hedonic Pricing Analysis

In hedonic pricing analysis, the implicit prices for the various qualitative and environmental characteristics were estimated by looking at the real markets in which that distinctiveness are effectively traded. Differences in these qualitative features of the land were thought to affect the flow of benefit from the property implicitly. Hence, to appraise the qualitative attributes of enhanced land use for agriculture and also to evaluate the improved values of the cropland, the data collected were subjected to hedonic pricing analysis. The results of the hedonic pricing estimates are represented in the Table 11.7. It could be noted that the coefficient of WQI was positive and significant at ten per cent probability level. It could be inferred that an increase in water quality resulted in significant positive change in the land value in the WTA. The distance of the sample farmer from the watershed structures showed a negative influence on land value but it was not significant. It could also be observed that proportion of cropland to the total land area had exerted a positive influence and was significant at one per cent level of probability. It could be discussed that for every one hectare increase in the proportion of cropland to total land holding, the land value increased by Rs 57,432

Table 11.7

Hedonic Pricing Estimates for WTA and NTA

Variables	*Coefficients for WTA*	*Coefficients for NTA*
Intercept	483740	265396.6*
	(17.745)	(11.388)
WQI	5086.89***	23065.18*
	(1.636)	(4.732)
PROX	–45.10	-24.05289*
	(–1.560)	(–4.515)
PCLA	57432.12*	14260.58
	(2.730)	(0.816)
WTDEPTH	–716.67	–1136.276*
	(–1.488)	(–3.245)
VAP	0.178146*	0.002028
	(4.222)	(0.056)

Figures in parentheses indicate the 't' value.

* Significant at 1 per cent level. ** Significant at 5 per cent level.

*** Significant at 10 per cent level.

keeping other variables constant. This variable had the expected sign. Similarly, the depth of water table showed a negative influence on the land value. However, this variable was not significant. As expected, the product value of the sample farm showed a positive relationship with land value and was significant at one per cent level of probability. Hence, it could be implied that for every unit increase in product value, the land value would increase by Rs 0. 17. Thus, from the above analysis, it could be concluded that the water quality, proportion of cropland to the total land parcel and product value showed a significant influence on the land value. Though, the distance and water table depth exerted a positive influence on the land value, the estimates of them were not significantly different at 5 per cent level.

The results also explain the hedonic pricing estimates for farms in NTA. Since the sample farms of NTA were located outside the geo-hydrological boundaries of the watershed, the sample farms did not experience any of the effects of the watershed. It could also be observed that the coefficient for water quality was positive and significant at one per cent probability level. It is implied that an increase in water quality resulted in increase in land value by Rs 23,065 in the NTA. The distance of the sample farmers from the watershed structures showed a negative influence on land value and was significant at one per cent level of probability.

The reason for this could be attributed to the higher implicit value that the sample farmers in the NTA assign to the proximity to watershed components. Thus, it could be inferred that for every one meter increase in distance from the watershed structures (major and minor check dams, percolation ponds) the land value decreased by Rs 24.05. The proportion of cropland to the total landholding though positive was not significant. It could be explained that for every hectare increase in the proportion of cropland to total landholding, the land value increased by Rs 14,261 keeping other variables constant. The depth of water table showed a negative influence on the land value and was significant to one per cent level of probability. It could be implied that for every one metre increase in depth of water table, the land value decreased by Rs 1,136 keeping other variables constant. Thus, from the above analysis, it could be concluded that the water quality, proportion of cropland to the total land holding and product value showed a significant influence on the land value.

Soil and Moisture Conservation Technologies: Cost Economics

The cost economics of the watershed components in the study area were analyzed to assess the returns on investment on soil and water conservation technologies. Hence, discounted financial measures such as Net Present Value (NPV), Benefit Cost Ratio (BCR) and Internal Rate of Return (IRR) were calculated. The results of the analysis are presented cropwise for different watershed technologies in Table 11.8.

Table 11.8

Cost Economics of Watershed Technologies

(discount rate – 10%)

Crops	*Maize*	*Chillies*	*Onion*	*Coconut*
Major Check Dam				
BCR	1.29	1.42	1.20	1.58
NPV (Rs)	9445	11426	8256	13677
IRR (%)	23	29	19	35
Minor Check Dam				
BCR	1.50	1.57	1.27	1.71
NPV (Rs)	10256	11048	7482	12633
IRR (%)	32	35	22	40
Percolation Pond				
BCR	1.54	1.52	1.08	1.63
NPV (Rs)	3382	13134	6794	3679
IRR (%)	32.24	32.12	14.3	37
Contour Bunding				
BCR	1.38	1.44	1.28	1.18
NPV (Rs)	915	994	796	678
IRR (%)	27	29	22	18.23
Land Levelling				
BCR	1.19	1.35	1.43	-
NPV (Rs)	976	1768	2164	-
IRR (%)	19	16	23	-

From the Table 11.8, it could be elaborated that the BCR ranged from 1.08 for percolation pond to 1.71 for minor check dams. The BCR for percolation ponds varied from 1.08 for onion to 1.63 for coconut. The range of BCR for contour bunding for the above crops varied from 1.18 for coconut to 1.44 for chillies. The BCR for major check dams though above unity were low ranging from 1.20 to 1.58 for onion and coconut. The reason for this could be attributed to the higher investment in the major check dams and the longer benefit realization period of the structure. The NPV of crops for the different conservation technologies reflect the impact of implementation of different conservation technologies. It is further noticed from the Table 11.8 that among the soil conservation technologies, major check dams had the highest NPV of Rs 13,677 followed by percolation pond with the NPV of Rs 13,134 for chillies. The major check dams had an NPV ranging from Rs 8,256 for onion to Rs 13,677 for coconut. Contour bunding and land levelling showed an NPV, ranging from Rs 678 for coconut to Rs 2,164 for onion.

The minor check dam had the highest IRR (40.12% for coconut). It was then followed by percolation pond with an IRR value of 37 per cent for coconut.

The major check dam gave an IRR value of 35 per cent for coconut. However, contour bunding and land levelling had the IRR values ranging from 16 to 29 per cent. The BCR, NPV and IRR values for coconut for conservation structures like major check dam and minor check dams were found to be high. The reason for this could be attributed to increased realization of benefits by the coconut crop under these structures. The overall results on feasibility analysis for the various soil and water conservation technologies, which include contour bunding, land levelling, minor check dams, major check dams and percolation ponds across the selected crops are encouraging and reiterated the advantages of the conservation structures in the watershed areas.

Garrett's Ranking

Garrett's ranking technique was applied to rank a set of benefits obtained from the impact of watershed development activities. The order of merit assigned by the sample respondents in the WTA was transmuted into scores using the formula given by

Garret and Woodworth (1977). The results of the Garrett scores obtained for different activities are tabulated in the Table 11.9 for easy understanding.

Table 11.9
Garrett's Score Value for the Benefits Obtained from Different Watershed Activities

Benefits	*Mean Garrett Score*
Improvement in soil and water conservation	78
Enhanced soil fertility	66
Change in cropping pattern	57
Increased cropping intensity	50
Yield enhancement	43
Groundwater recharge	34
Resource conservation	22

It could be inferred from the above table that among the benefits listed, the highest score was recorded for soil and water conservation (78). This, in turn, had indicated that the construction of different water conservation measures such as percolation ponds and check dams have greater impacts on the conservation of water and controlling of soil erosion. This benefit was then followed by the improvement in soil fertility with a score of 66. The results of the table also ranked the other benefits like cropping pattern, cropping intensity, yield increase, groundwater recharge and resource conservation with Garrett's scores of 57, 50, 43, 34, and 22, respectively.

Conclusion

As reported already, in India, about 70 per cent of arable land is rainfed and depends on monsoon rains and around 45 per cent of these drylands still continue to depend on natural precipitation even after realizing the full irrigation potential. It is also understood that the main source of instability in the overall agricultural production in India is due to unevenness in the production scenario of dryland agriculture. This dreary picture of dryland agriculture is a major cause of concern because it is the natural habitat of most of the crops, India's rainfed agricultural sector also provides livelihoods for hundreds of millions of people and it is the source of nearly half of the value of the country's agricultural production. In order to conserve the precious dryland natural resources, viz., soil, water etc. and also to sustain the existing resource endowments for future entitlement, the governments (both central and state governments) have formulated 'watershed approaches' in eighties. As unexploited irrigation potential is increasingly scarce, planners look increasingly to rainfed agriculture to contribute to food basket and economic development in the decades ahead.

Under the present state of affairs, research targeted to rainfed areas should be accompanied by increased investment in rural infrastructure and policies to bridge the gap between potential yields in rainfed areas and the actual yields achieved by farmers. Important policies for achieving the above propositions include development of irrigation potentials through appropriate conservation mechanisms, extending higher priority for rainfed areas in agricultural extension services and access to markets, credit, input supplies etc. Successful development of rainfed areas is likely to be more complex than in high-potential irrigated areas because of their relative lack of access to infrastructure and markets, and their variable agro-climatic environments. Investment in rainfed areas, policy reform, and transfer of technology such as water harvesting, conservation and sustainable resource use will therefore require stronger partnerships between agricultural researchers and other agents of change, including local organizations, farmers, community leaders, NGOs, policy makers and supporting agencies. The conservation and sustenance of precious natural resources such as soil, water etc. in dryland areas through appropriate water conservation mechanisms for future entitlement deserves immediate attention. The field level experiences also indicated that the cropping intensity of the sample farms in watershed treated area (WTA) was higher. The productivity of major crops also exhibited enhanced amount. The soil and moisture conservation works under watersheds have also extended the beneficial effects on recharge of groundwater, combating soil erosion and change of cropping pattern with positive effects on the productivity of the crops. Hence, watershed activities may be scaled up and implemented in drylands on large scale to reap the real benefits of' watershed activities, sustainability of natural resource base and strengthening of economic prosperity of our country.

■ References

Alagumam, T., 1991, "An Economic Evaluation of Soil Conservation in Avinasi Taluk. Coimbatore District, Tamil Nadu", *Indian Journal of Agricultural Economics.* 46(3): 300-301.

Cai, X, and M.W. Rosegrant, 1999, "Irrigated and rainfed crop area and yield in China, India, and the US", International Food Policy Research Institute, Washington, D.C. (Mimeo).

Chaurasia, S.P.R. and L.R. Singh, 1991. "Optimal Cropping Pattern for Minimizing Soil Loss in Nauraar Watershed of Almora District in Uttar Pradesh Parameters", *Indian Journal of Agricultural Economics,* 46(3): 320.

Gittinger J. Price. 1972, *Economic Analysis of Agricultural Projects,* John Hopkins University Press, London, p. 64.

Gupta, S.B., Rajendra Prasad and R.K. Pandey, 1973, "An Economic Evaluation of Soil Conservation Measures in Varanasi District, Uttar Pradesh", *Indian Journal of Agricultural Economics,* 28(4): 205-211.

Hofwegen P. Van and M. Swendsen. 2000. "A Vision of Water for Food and Rural Development. World Water Vision Report, Paris", United Nations Educational Scientific and Cultural Organization (UNESCO).

Lal, B., 2001, "Impact of National Watershed Development Project on Rainfed Agriculture in Alwar District of Rajasthan", *Agricultural Science Digest,* 21(3): 157-160.

Mahandule, D.K., R. Pawar, D.L. Sale and S.A. Kadam, 1991, "Effects of Watershed Development Programme on Resource Use Structure and Returns of Farms in the Drought Prone Areas of Western Maharashtra", *Indian Journal of Agricultural Economics,* 46(3)287

Rosegrant, M.W., and X. Car, 2000, "Modeling Water Availability and Food security a Global Perspective: the IMPACT Water Model", A paper presented at the United Nations Global Modeling Forum in New York. February 2-4.

Sekar, C., 2001, "Externality Effects of Common Property Resource Degradation", *Indian Journal of Agricultural Economics,* 56(3)346-357.

Sekar, C. and Senthilnathan, S., 2004, "Comparative Evaluation of WTA (Watershed Treated Area) and NTA (Non Treated Area) in Coimbatore District of Tamil Nadu", Research Report-2004, Department of Agricultural Economics. Tamil Nadu Agricultural University, Coimbatore.

Sinden. J.A., 1974, "A Utility Approach to the Valuation of Recreational and Aesthetic Experiences", *American Journal of Agricultural Economics.* 56: 61-72.

Government of India, Planning Commission, 2001, World Bank Report, 1999: www.dhan.org; www.fao.org; www.ifpri.ogg; www.tn.gov.in

Chapter 12

Watershed Management in India

Brief Comment on an ICRISAT/IWMI Report*

Rakshat Hooja

'WATERSHED development' is a buzz phrase in India these days and is being looked upon as one of the most viable and holistic paradigms for the development of rainfed regions of India. What began as ad hoc soil and water conservation measures over three decades ago has evolved into an integrated, inter-sectoral community-based approach that tries to put people's livelihoods at the heart of the debate on natural resource management, and the Indian government as well as the partnering non-governmental organizations (NGOs) have

---

* A comment on P.K. Joshi, Vasudha Pangare, B. Shiferaw, S.P. Wani, J. Bouma and C. Scott, *Socio-economic and Policy Research on Watershed Management in India: Synthesis of Past Experiences and Needs for Future Research,* Global Theme on Agrosystems, Report No. 7, 2004, ICRISAT, ADB, IWMI, Patencheru, Andhra Pradesh.

adopted this integrated watershed management approach as a vehicle for sustainable agricultural transformation and rural livelihood security in rainfed areas.

The report under consideration is based on a study undertaken by the International Crop Research Institute for the Semi-Arid Tropics (ICRISAT) and the International Water Management Institute (IWMI) with the objective of synthesizing the knowledge gained by the adoption of the watershed development approach in India and identifying needs for future research. The report has attempted to look at policy and socio-economic aspects of the watershed development paradigm and along with analysis of national level policies related to land, water, forest, watershed development has also made use of case studies of both model watersheds and watershed development schemes/approaches.

With the authorial credit being shared by six people, namely, P.K. Joshi from the International Food Policy Research Institute (IFPRI), New Delhi, Vasudha Pangare from World Water Institute, Pune, B. Shiferaw and S.P. Wani from ICRISAT, Andhra Pradesh and J. Bouma and C. Scott from IWMI, South Asia Office Patencheru, the influence of the different subject specializations of the authors is clearly felt. However, overall, the work under review would best fit into the category/domain of what is nowadays being called "ecological economics". Written in a descriptive style, though quantitative data in the form of tables is also provided in some cases, the focus seems to be to get a qualitative and explanatory understanding of the situation and to identify the spheres that require more research.

The report starts of with a review of the theories, concepts and institutional issues associated with watershed management, followed by a presentation on the policies that have a link with watershed development in India. The section on theories is extremely well written and in the small amount of space devoted to it is able to successfully touch upon, elaborate and explain complex concepts like externalities in watershed management, common pool resource management, institutional alternative, etc.

For example, a very simple definition of the term 'externalities' (this is a term that, as the author has experienced while attending seminars and conferences related to natural resource management, creates a lot of confusion and is used to denote different things by different people) has been provided as: "The unintended spillover effects of private resource use decisions that affect production or consumption of farmers other than the resource owner and are not mediated through the market mechanism are commonly referred to as externalities."

Similarly, while looking at common pool resource management, the report examines the views of a number of scholars including Elinor Ostrom and Anil Agarwal and finally distinguishes four categories of factors that play an important role, viz., resource system characteristics, user group characteristics, institutional arrangements and external environment.

One important theoretical aspect of watershed development that this section only looks at peripherally is that of participation. While going through the volume under review, it may have been useful for the reader to also get to know that at least one typology of participation in watershed projects has already been adapted and presented in the book edited by Farrington, Turton and James titled *Participatory Watershed Development: Challenges in the Twenty First Century* (1999) in which one of the authors of the current work, Vasudha Pangare, was also a contributor. The typology is reproduced below:

Typology	*Components of Each Type*
Passive participation	People participate by being told what is going to happen or has already happened.
Participation in information giving	People participate by giving answers to questions posed by extractive researchers and project managers.
Participation by consultation	People participate by being consulted and external agencies listen to their views. External agencies define both problems and solutions.
Participation for material resources	People participate by providing resources, for example, labour in return for cash and food.
Functional participation	People participate by forming groups to meet predetermined objectives relating to the project, which can involve the development or promotion of externally initiated social organization.
Interactive participation	People participate in joint analysis, which leads to joint-action plans and formation of new groups and strengthening of old ones.
Self-mobilization	People participate by taking initiatives independent of external change systems.

The section covering important policies and guidelines related to watershed development including the National Agricultural Policy, Water Policy, Land Policy and Forest Policy provides important background information to readers not familiar with the Indian scenario but otherwise, due to lack of space and more focus on the case studies being presented in the next section, does not go into too much detail. In fact, this section is very dull and reads like a government report and actually quotes a lot from the actual policy documents which it summarizes and superficially examines. Also macro-level examination of policies related to issues like administration, role of panchayats, rural development programmes, infrastructure development etc. is not attempted by the authors and it is only when the discussion shifts to the Watershed Development Guidelines that attention is once again paid to detail.

The common guidelines of 1994 and 2000 issued by the Government of India have been analyzed in the report. The special focus of the analysis is on the institutional arrangements and a very clear and simple graphic depicting the organizational and institutional arrangements, from the national to the state, district, project and micro-watershed level, as visualized in the 2000 common guidelines, is part of the report and will prove to be useful to practitioners and researchers of watershed development alike. This section also borrows some material from the DFID, International Institute of Rural Reconstruction (IIRR) and OIKOS publication titled *Social and Institutional Issues in Watershed Management in India (2000)*, which is one of the best handbooks/manuals on the subject that this winter has seen till date. Readers may like to compare this section on the common guidelines with this observers summary and views on the common guidelines of 1994, 2000 and 2001 available in two co-authored articles "Community Based Organizations and Panchayats as Instruments of Governance in the Sphere of Watershed Development", *Prashasnika*, Vol. 32, No. 1, January-June 2005 and in "Changing Role of Panchayat Bodies in Watershed Development" in Y.S. Sisodia (ed.), *Functioning of Panchayati Raj System*, (2005) where it has been brought out that these common guidelines were better than the subsequent Haryali Guidelines of Government of India of 2003 in many respects.

Strangely enough, the section on the history of watershed management in India is placed following the analysis of the 1994 and 2000 watershed guidelines while it would obviously have fitted in better earlier in the report.

The bulk of the report consists of six case studies, which have been selected on the basis of being representative of the functioning, processes and approach of the different watershed programmes in the country and which are all considered relevant for the country in today's context. The six approaches that have been studied are those of the Adarsh Gaon Yojana of the Maharashtra Government; the Rajiv Gandhi Watershed Mission funded by the Central Government and implemented by state Government; Mysore Resettlement and Development Agency (MYRADA), initiated and implemented by that and other NGOs; Sukhomajri Model implemented by the Indian Council for Agricultural Research (ICAR), State Government and local level institutions with funding from Ford Foundation; the ICAR model; and the Consortium model developed by ICRISAT. Each case study gives the background as well as the unique approach to watershed development of the particular programme, the institutional arrangements, and the impact of the watershed development before moving on to description and analysis of specific watersheds and finally to lessons learned from the experiences. Incidentally, almost all the approaches have been written about extensively; for example, the MYRADA approach is described by its initiators in detail in *Resource Management in Rainfall Drylands: An Information Kit* (MYRADA, IIRR 1997).

Going by merely a reading of the six approaches in this report the author felt that the ICRISAT approach appears to be the best. But it is the one with the least field implementation examples at present.

This observer has had a chance to visit the Kothapally village watershed (as part of a seminar field visit led by Dr S.P. Wani, one of the co-authors of the report under review) in Andhra Pradesh in February 2005. This watershed is the example looked at in the case study on the consortium model propagated by ICRISAT and one can vouch that in this case the study accurately depicts the ground reality. As the case study has mentioned the focus or emphasis on individual farmer-based conservation measures for raising productivity in individual farms along with community-based soil and water conservation measures is one of the reasons for the high level of participation by the villagers and has contributed to the success of the watershed. Also, the technology and technological back stopping by ICRISAT is another factor for the success of the Kothapally watershed and during the visit one could see that the main benefits were coming from better cropping techniques and use of improved and more suitable varieties of crops by the farmers (an area where the core competence of ICRISAT lies) and even successful plantation of bio-diesel crops in the less fertile soil areas of the watershed had taken place. *The Hindu* newspaper of October 9, 2005 has, in an article titled "It is Plenty Amid Scarcity in Kothapally" mentioned that the Andhra Pradesh government has selected the Kothapally model for its Drought-Prone Areas Programme and the Andhra Pradesh Rural Livelihood Programme. It now needs to be seen if the success of this model will withstand the test of scaling up.

The authors of the report being commented upon have, in the end, compared the different case studies using qualitative ordinal scales along a few major components of a holistic and integrated approach to watershed management. Nothing interesting here except for the fact that on the question of creation of non-farm employment five of the approaches are given a low rating and the sixth, i.e., MYRADA, is given a 'limited' rating (because of the presence of self-help affinity groups/credit groups). It seems that this should be an area for future focus in holistic watershed programmes and should be made a part of the two-page bullet-pointed section right at the end of the report on knowledge gaps and suggested areas for future research.

Overall, this eighty-page report is one of the best introductions that a layperson can have to the watershed development concept and activities in India. Almost all the relevant information is briefly available compiled at one place. Experienced practitioners will also find it interesting to read and useful to keep as a reference document.

Chapter 13

Managing Water Supply in the 21st Century Metropolis

Lessons from Delhi

Rakshat Hooja and Rakesh Hooja

ONE area where the emphasis of the modern world is somewhat different from the pre-industrial era is the need to make safe drinking water and sanitation available in copious quantities for large urban clusters and the mega-cities, which are fast growing in numbers and size.

As urban centres increasingly grew across the world during the 19th and 20th centuries, the European idea of public-funded authorities managing municipal water supply and sewerage found acceptance everywhere, with the costs of these being met by the concerned governments via cross subsidies, and other available monetary resources since water was viewed as a public good and the requirement of safe drinking water as a basic need was recognized. But, today, arguments are being put forward supporting the need to price water and free the government of the responsibility of ensuring safe and adequate water supply.

The problems associated with urban water supply have increased substantially during the last fifty years. Among other things, for the first time in history, half of the world's population is living in towns and cities. According to Uitto and Biswas (2000: back cover):

> By the year 2025, it is estimated that more than two-thirds will be urban dwellers. While the fastest growth of cities is taking place in the developing world, urbanization is a global phenomenon, closely related to environmental issues. The rapid growth of urban centres will place tremendous stress on the environment and pose formidable problems of social and institutional change, infrastructure development, and pollution control. Water will be one of the key resources for sustainable urban development. It is needed for virtually every human endeavour—for household use, agriculture, industry, leisure - and water also has an important ecosystem function. Provision of sufficient water and preventing pollution however, are formidable tasks.

Because of the rapid growth of world population, mass consumption and pollution, among other factors, the availability of drinking water per capita is shrinking. It is believed that 40 per cent of the world's inhabitants currently have insufficient fresh water for minimal hygiene.

Modern urban water supply systems get water from a variety of locations, including aquifers, lakes, rivers, wells, desalinated seawater, and other sources. The water is then purified. The intake from these water sources usually is through a large cage-like box designed to screen out large particulate matter before it enters the system. After it is sucked in by a pumping station or allowed in by a gravity-feed system, it is usually filtered further, chlorinated, and then pumped either to holding locations like water towers or reservoirs, or fed directly into the user's spigot.

Municipalities typically run water supply systems, although sometimes this is the job of a regional supplier or water utility that has an independent governmental structure and taxing authority, for example, the Delhi Jal Board, or DJB. Once water is used, it has to go somewhere. Typically, waste water is piped away in a sewer system, which is again almost always a service provided by the same authority as the water supply, since usage of one system implies usage of the other. Of late, this established system of water delivery and disposal is under threat from informal markets as well as the move towards looking at water as a commodity, with its associated aspect of privatization of water delivery mechanisms. The problem has arisen because of governments finding it difficult to fullfil satisfactorily the basic water needs of their citizens.

Full-fledged privatization of water companies has however been contested on several occasions, on account of bad quality of water, increasing prices, etc. An often mentioned example is that of Bolivia, where the proposed privatization of water companies was met by popular protest in the city of Cochabmba in 2000, and which eventually led to the ouster of Bechtel, an American engineering firm, which had been planning to take over and run the water company.

De facto Privatization

While this debate over the pros and cons of government versus private supply of water is still open, a trend that is increasingly being noticed is that from the consumer/citizen's viewpoint, there exists a situation of 'de facto privatization' because the consumer is anyway having to pay more to get safe drinking water, than what is being charged by the utility. This extra charge (or 'de facto privatization', the term is being used because the income from this is going into private hands while in theory governments are still providing safe water at subsidized costs) may for example be in the form of

1. Corruption, i.e., illegal extra money being given to municipal/government functionaries to get access to water.
2. Citizens needing to instal extra equipment to even get their fair share of water. These include needing to instal booster pumps, extra underground or overhead tanks and even water filtration or purification devices at home as in many cases the government/municipal supply is not safe to drink.
3. Needing to buy bottled water at a premium price.
4. Getting water delivered via alternate private delivery mechanism such as water tankers as the government supply is inadequate.

It may be argued that that the informal private markets for water are growing at a much faster rate than the increase in municipal supply of water.

India and Delhi

Before moving on to Delhi, let us briefly look at the Indian scenario. The nation has more than 250 million city-dwellers. Experts predict that this number will rise even further, and by 2020, about 50 per cent of India's population will be living in cities. Already India, as per the 2001 census, has 35 cities with million plus population and this number will grow putting further pressure on the already strained centralized water supply systems of urban areas. According to an uncredited document posted at www.rainwaterharvesting.org/Crisis/Urbanwater-scenario.htm:

> The urban water supply and sanitation sector in the country is suffering from inadequate levels of service, an increasing demand-supply gap, poor sanitary conditions and deteriorating financial and technical performance.
>
> According to Central Public Health Engineering Organisation (CPHEO) estimates, as on 31 March 2000, 88 per cent of urban population had access to a potable water supply. But this supply is highly erratic and unreliable. Transmission and distribution networks are old and poorly maintained, and generally of a poor quality. Consequently, physical losses are typically high, ranging from 25 to over 50 per cent. Low pressures and intermittent supplies allow back siphoning, which results in contamination of water in the distribution network. Water is typically available for only 2-8 hours a day in most Indian cities. The situation is even worse in summer when water is available only for a few minutes, sometimes not at all.

It may be noted that Paris with a lower per capita daily quantity of water than many Indian cities manages to provide 24 hours a day a water supply thus indicating that along with quantity, it is a matter of proper management to improve water supply.

The urban water supply systems in most cities depend on surface water sources like rivers and lakes. Water from a distance of 200 km is brought into Chennai whereas Bangalore gets its water from the Cauvery river, which is 95 km away. Where surface water sources fail to meet the rising demand, groundwater reserves are being tapped.

In the case of Delhi there seems to be a situation of constant water crisis, more so during the dry season, when the situation gets particularly worse. As the demand-supply gap widens (see Figure 13.1), more groundwater is being exploited. Of the water supplied by the municipality, approximately 11 per cent comes from groundwater reserves and remaining from the Yamuna river. It is, however, difficult to determine the total quantity of groundwater extracted because a large number of tube-wells (owned by individuals, industries and bottled water companies) remain unregistered.

Figure 13.1
Demand and Supply Gap in Delhi[2]

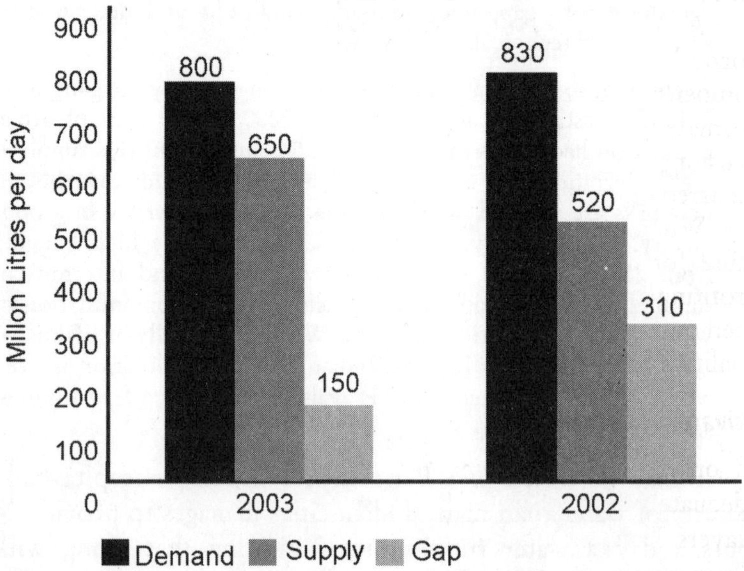

Source: Delhi Jal Board.

The pattern of water consumption shows that the share of domestic water consumption increased from 75 per cent in 1976-77 to 88 per cent in 1995-96. Per capita consumption of water per day increased from 32.07 gallons during 1980-81 to 49.03 gallons in 1990-91 but dropped to 43.76 gallons in 1995-96. The per capita consumption was 43 gpcd as on 31.3.99 (Economic Survey of Delhi, 2000).

The population figures released by the Registrar General of India indicate that the population of Delhi was around 134 lakh in March 1999. Based on a norm of 70 gallon per capita per day prescribed by Master Plan of Delhi 2001, the water requirement for this population in 1999-2000 would be 939 million gallon daily (MGD). So, the actual gap might be even higher than the Delhi Jal Board figures suggest.

In fact the claims that there is adequate water in Delhi are not echoed when one looks at the problem from the consumers' side. It is not uncommon for residents of colonies like Vasant Kunj to pay over hundred rupees for 1,000 litres of water to private water tanker operators as the Delhi Jal Board Supply in the area is minimal (Agenda 2005).

According to a study (Lorente and Zerah, 2003), the total cost borne by Delhi households on account of strategies devised to compensate water shortages like pumps, tankers, storage, is estimated to be 6.5 times higher than what they pay for their consumption of water. This figure does not include the cost incurred on water purification devices.

When looking at Delhi, two reasons may be identified as the cause for a number of private players getting involved in and profiting from the water supply business. The first has already been outlined above as the scarcity of raw water. The second is the quality of water supply.

Private Supply via Water Tankers

In order to meet the needs of residents who are not getting adequate supply of water from the Delhi Jal Board, many private players have set up businesses of supplying water on demand. Their operations are simple. They would normally own/have a contract with a ground water extraction unit (borewell) at the outskirts of the city or in one of the many villages that now fall within Delhi. Everyday they would fill up their fleet of three or four tankers and station them near colonies where the DJB supply has problems. The drivers and helpers are provided cell phones. Whenever a customer needs water he/she rings up the cellphone number of the supplier and within half an hour the water tanker is at their door step to fill up their water tank. The rates vary in different areas but generally fall within the range of Rs 80-120 per 1,000 litres. The tankers have a capacity ranging between 50-60 thousand litres of water and are thus able to earn daily cash revenue of around Rs 5,000. Even after accounting for the costs the business does seem very profitable.

The Delhi Jal Board also has a scheme of providing free tanker water delivery in areas where piped water delivery has been affected for some reason but even after the residents' complaint there is no guarantee when and if the tanker will come after their complaint and as water is essential for their daily needs they prefer private suppliers even if they have to pay higher amounts.

The Water Purifier Business

The water being supplied by the utility is supposed to be drinkable as it has undergone treatment. In reality anyone who can afford it makes sure that he/she has a private water purifier in their house and only water that has been treated by their drinking water purifier is used for drinking. There are a number of reasons (like lack of confidence in the utility supply, savvy marketing campaigns by manufacturers of filtration units etc.) why this is happening in Delhi (and most large urban centres of India) but the result has been that a multi crore personal water purification business has developed in India. And the money that is fueling this industry is the money that consumers are paying for getting drinking water in their homes. In effect once again there is a premium being paid for drinking water that is going into the hands of private players. For example if one has bought a purification unit for Rs 6,000, which lasts for three years, during which time one also spends Rs 500 per year for new filters, one ends up paying Rs 7,500 extra over three years or Rs 2,500 per year extra. So the net result is that the user is paying over Rs 200 per month to a private company for his/her drinking water supply.

Related to the purification business is the bottled water market that is still developing in India. The Delhi Jal Board has itself launched a bottled water brand called "JAL" and most of the global majors are now present in India.

Resident Welfare Associations (RWAs) and Water Supply in Delhi[3]

In most areas of Delhi RWAs have become important social institutions that play an increasingly significant role in the lives of the residents of these areas. Over time most of the RWAs have been registered under the Societies Act. The Government of Delhi has also launched a "Bhagidari" scheme where the authorities form partnerships with the local RWAs for carrying out the many activities.

An independent study[4] concluded that the single most important reason for Resident Welfare Associations in South Delhi remaining active and influential is increasing urban stakeholder activism with regards to civic amenities especially water supply among the middle and upper classes. The table below shows that in certain areas of Delhi like Vasant Kunj few residents seem to be satisfied with the water supply.

Figure 13.2

Percentage Satisfied with Water Supply in the Colony/residential Area

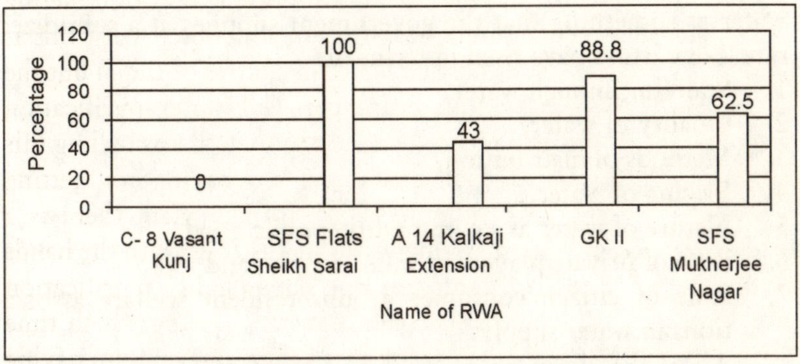

The study also noted that in many areas special schemes to ensure water supply and maintain water costs had been initiated by the RWAs. For example in Vasant Kunj the RWAs have tried to bargain with the private tanker suppliers to fix a lower rate, and have taken up home delivery of the Delhi Jal Board mineral water "JAL" that costs half of what the private suppliers are charging, and have had agreements with the DJB so that free tanker supply is available when piped water supply fails etc.

It is also interesting to note that in areas where the RWAs have existed for a longer period with many long time residents as members such as GK II and Sheikh Sarai the problems seem to be less.

Another point of interest is that not everyone is happy with the functioning and benefits of the RWAs. The following statement extracted from an interview of K. Gulati in Vasant Kunj seems to represent the views of a number of respondents across four of the RWAs (In GK II 100% of the respondents thought that that the RWA was democratic and were happy with its performance):

> "The benefits have reached only few people, but we have been able to voice our difficulties through the Resident Welfare Association. The RWA has not been of any help in improving water supply except for areas where elected members live."

The Future

The above examples show that the very nature of urban water supply is changing in Delhi. It is no longer possible to look at water as something that the government supplies at a subsidized rate. New issues have risen in terms of

1. Sourcing enough water,
2. Quality of water,
3. Methods of distribution,
4. Pricing of water,
5. Nature of water as a commodity/public good,
6. Role of private players in water supply, and
7. Role of citizen/consumer groups/resident welfare associations in water supply.

All of the above points need to be addressed as one tries to formulate the new paradigm for water supply in the 21st century metropolis, especially in rapidly developing countries like India. The stress on the government infrastructure is invariably leading to supply problems and shortages. And this is where private companies have stepped in. Water is being treated as a commodity and the normal demand supply curve now applies to it. The government, municipality and the Delhi Jal Board needs to keep this reality in mind as they plan their future actions and policies.

Recommendations for Managing Water in Indian Metropolis and Mega-Cities

The management of urban water supply should have as its planning and implementation unit not the main city habitation or municipal area but the city and its suburbs and attached peripheral semi urban and rural areas. Thus, water bodies being developed or restored in the proximate areas to the city, large water treatment plants, attached industrial and agricultural-horticultural use areas, and planning for the expansion of the city would all be part of the larger urban agglomeration that could be managed by a special water utility and not by the multipurpose municipal body covering a smaller area. The municipality could thus be freed from the problem of water supply and the water utility could develop expertise in the same.

For augmenting water supply the largish water bodies, dams being developed/restored on the peripheries of the urban agglomeration need to be supplemented by rain water harvesting on a community/mohalla/RWA level rather than individual level. Rainwater harvesting should be aimed at recharging ground water as well having some local storage for use during emergencies.

Similarly, for water quality the large treatment plants on the outskirts of the city could be supplemented by smaller more efficient treatment plants at intermediate points in the supply system. This would enable effluent treatment plants for industrial areas and sewage treatment plants to be managed separately, as well as allow ground water to be drawn and treated in a decentralized manner.

The specialized public water utility should manage the main storage, treatment and supply system, but water should be provided in bulk to community-based organizations (CBOs) or resident welfare associations, especially in middle and high income areas wherever they are capable and willing to take up this responsibility. The rules for this should be enabling and not mandatory, with utilities and municipalities motivating RWAs/CBOs to take up the task. The utility should also undertake capacity building exercises in terms of providing training and technical advice so that the community could be empowered to undertake the final mile of water delivery as also technically sound rain water harvesting, controlled use of groundwater and management of *mohalla* drains. The utility would require specialized staff to deal with the RWAs/CBOs; staff which understands the internal dynamics of RWAs. This concept is similar to the Participatory Irrigation Management (PIM) concept in irrigation. The benefits would be that the utility would deal, through a zonal system, with a few hundred RWAs or community organizations rather than thousands of customers and thus be able to concentrate on reducing water losses in the main and sub-main distribution network. The RWA/CBO could choose its water distribution model and internal rates. Of course there would be a need for a regulatory or arbitration body to oversee the agreements between the utility and the RWA/CBO. Public subsidies would need to be made available in cash and in terms of not charging for bulk water supplied during the initial period to the RWA/CBO for it to put its new system in place or to undertake deferred maintenance if utility does not do so before handing over system to RWA. In fact the final water treatment should also be at the RWA level either by a group of RWAs using a common plant run by RWAs, or the utility, or privately.

The utility would need to provide free water to poor/slum areas in the city as well as maintain the sewerage system. This may be managed by providing a cash subsidy to the utility which is made up by adding a cess to the house tax being collected by the government/municipality.

Waste water recycling needs to be introduced in a big way. Precious drinking water should not be wasted in uses like car washing, maintaining lawns, cooling and in industries etc. This would entail two sets of water supply with differing charges - one for cooking, drinking and domestic use and the second of recycled water for other uses.

Reduction of conveyance loss and checking of unaccounted for water would need to be high on the agendas of both water utility and CBO/RWA as this water would become available for supplying. In fact incentives need to be provided to both the provider and the consumer (RWA and individual resident) to conserve water. This incentive could be in terms of differential pricing linked to water use or cash subsidy.

The current approach of the municipality/large utility supplying water to every resident may not remain feasible as the mega-cities of India continue to grow. "Out of the box thinking" is essential for both the demand side and supply side management of water delivery. New roles need to be defined for the utility, community as well as private players in order to make sure that the problems of water supply are effectively solved in the Indian mega-cities of the 21st century.

Notes

1. The ideas expressed here are personal and academic and do not represent the organizations where the authors are currently working.
2. The Economic Survey of Delhi 2000 and the Delhi Jal Board have used different norms. Delhi Jal Board has estimated that if the norm of 60 gpcd for planned colonies, 34 gpcd for Regularized-unauthorized colonies and 11 gpcd for other areas, is adopted, the water requirement for the present population of Delhi will be about 440 MGD as against present capacity of 591 MGD. Following these norms, water requirement for the projected 22 million population of Delhi in 2021 will be about 747 MGD which can be met through availability of 300 cusecs (140 MGD) of raw water from Tehri Dam. Ministry of Urban Development, Government of India has also recently announced the water supply norm of 37 gpcd for metro cities which is lower than 43 gpcd availability of water at present in Delhi.
3. This section is based on a study undertaken by Rakshat Hooja as an Independent Fellow Project for Sarai/CSDS, New Delhi in 2006.
4. The study consisted of detailed data collection from 5 RWAs of Delhi in an effort to examine the RWA as a viable group for urban stakeholder activism.

References

Agrawal, Anil, Sunita Narain, and Indira Khurana. 2000. *Making Water Everybody's Business*, Centre for Science & Environment (CSE), New Delhi.

Agenda. 2005. Special Issue on 'Politics of Water', *Agenda*, Issue 3, October 2005.

Asian Development Bank 1994. *Managing Water Resources to Meet Megacity Needs*, Asian Development Bank, Manila, Phillipines.

Ball, Philip. 1999. *H2O, A Biography of Water.* Wiedenfeld & Nicolson, London.

Black, Maggie. 2005. *No Nonsense Guide to Water*, Rawat Publications, Jaipur & New Delhi [originally published 2004, by The New Internationalist Publication, Oxford].

Black, Maggie, with Philip Talbot. 2005. *Water a Matter of Life and Death: Water Supply and Sanitation in Village India.* Oxford University Press, New Delhi.

Chowdhary, Pooja. 2006. "Managing Drinking Water in Indian Cities: Comparative Perspective" in P.N. Rao (ed.), *Urban Governance and Management: Indian Initiatives,* IIPA and Kanishka, New Delhi.

Frontline. 2006. 'Thirst for Profit' (pp.4-9), 'Private Water, Public Misery' (pp. 10-21), and 'Bottled Loot' (pp. 23-25), & other articles in *Frontline*, April 21, 2006 issue.

Llorente, M. and M.H. Zerah, 2003. "The Urban Water Sector: Formal Versus Informal Supplies in India", *Urban India*, Vol. 22, No. 1.

Malick, Nidhi. 2006. "Management of Water Supply: A Case Study of Delhi", *Ashwatha*, Vol. 6, No. 2.

National Capital Region Planning Board (India). 1999. *Economic Survey of Delhi.* National Capital Region Planning Board, New Delhi.

Uitto, J.I. and Biswas, A.K. 2000. *Water for Urban Areas.* United Nations University Press.

UNESCO. 2003. *World Water Development Report.* (WWDR), UNESCO, Paris.

USAID – India 2006. Papers of Seminar on Transforming Water and Wastewater Services in Urban India – Lessons of International Best Practices, Jaipur, August 2006.

World Bank. 2001. World Bank Background Paper – International Conference on New Perspectives on Water for Urban & Rural India – 18-19 September, 2001, New Delhi, World Bank.

Chapter 14

Water Resources and Interlinking of Rivers

Issues Concerning Inequalities, State Intervention and Sustainable Development

D.K. Verma and A. Sohrot

Water, the essence and sustenance of life, is among the biggest and the most crucial natural resources both in terms of quantity and quality. Disputes on water are on increase day by day and water is becoming an issue of concern and a report of UN speculates that it may even lead to war in the coming years. The continuing drought spells and scarcity of water in many parts of our country even in those areas where ample water sources and reservoirs are located, emphasize the need to appreciate water and its conservation as a valuable resource. As the population increases, the need for water availability and accessibility will continue to grow dramatically and will exert additional stress on our limited resources exacerbated by depletion and pollution. In order to meet the needs of existing and future populations (sustainable development) and to ensure that habitats and ecosystems are protected (ecological balance), the water resources must be conserved for the society in general (environmental economics) and weaker sections in particular (social welfare). Sound resource management philosophy together with a careful policy, efficient planning for use of water is essential for ecological balance.

Efficient use of natural resources and their conservation can have major environmental, public health, and economic benefits by helping to increase resource availability and accessibility to the needy people. Water conservation, particularly in resource scarce regions, is one way of marching towards sustainable development. However, a reorientation shall be required in all sectors related to water resources and its consumption from sustainable development perspective, i.e., philosophy, policy, planning and people's participation in management and conservation of water resources. Dr B.R. Ambedkar was a great visionary of modern India. His thoughts on water and its conservation are of much relevance for the sustainable development of water resources. Particularly, Dr Ambedkar's views on interlinking of rivers and multipurpose use of water resources have become more concerning and relevant.

Important and urgent consideration is needed for ground water resources. Nobody pays for the underground water. Ownership of surface land does not provide ownership of depth of the owned land including underneath resources. Land price is paid in square units for length and breadth and not in cube units to include depth also. There should be an important consideration in national water policy in terms of favouring environmental economics giving more monetary value to environmental conservation than commercial use. This paper reviews the current policy and planning with regard to water resources and its conservation and also attempts to bring forth a framework based on environmental ethics to conserve the available water resources to meet the needs of the present generations and also enabling the future generations to meet their own needs and more particularly of the weaker sections of our society who are in a greater need of quality resources than ever.

Water Resources: Geographical Inequalities and Unsustainable Practices

Depletion of water resources is a serious concern. However, more concerning is the socio-economic and geographical inequalities in availability and accessibility to water which is a greater concern of sustainable development. Geographical inequalities of natural resources further marginalize socio-economically backward classes. Quality and quantity of land and water resources are directly linked to the socio-economic status of people. Even during the normal period also water is available and accessible to weaker sections poorly both in terms of quantity and quality. The situation becomes worse during the scarcity period and drought.

Physical scarcity of water is an ecological concern. However, problems related to available water in terms of its accessibility to people equally is more of social nature due to overconsumption by middle and upper classes, improper administration, management and conservation of water resources that further marginalizes weaker sections. The availability of water is important but more important is its accessibility in equal terms to all sections of the society. Water scarcity is more due to overconsumption, other than intended use (abuses) of water and corrupt practices in its management. Therefore, the current policy, planning and programmes need to be reoriented towards meeting the needs of the present generation particularly those of weaker sections. A few case studies are discussed here to highlight the unsustainable practices in accessibility, consumption, conservation and management of water resources.

Case 1: Drinking Water Supply in a Small Town

The drinking water supply in a town 'X', is presented as a case study because the population of town is less than one lach, it has a local administrative body that can manage water from multiple sources. It could be considered an ideal township both in terms of population, water availability and supply management. The case is selected purposively because the discussion point is resource availability, accessibility and its consumption.

The total population of the town (as per 2001 census) is 70,834 living in 3,400 authorized households. However, as per the average family size of five the total number of households could be around 14,166. In other words there may be about 10,766 unauthorized dwelling units in the town.

The total daily supply of the drinking water is met through a supply (18,71,800 litres) pumped from a river at a distance of more than 50 kms and a local tank reservoir (5,86,800 litres) located at distance of 5 kms. On record, there are about 50 tubewells, 200 pumps and 35 open wells which supply 16,06,000 litres water daily. In all the total daily water supply in the town is 40,64,600 litres. As per the official estimates, the per capita drinking water supply in the summer season in the town comes to a little over 57 litres per day. The local administrative body makes a payment of approximately Rs 82 lakhs per month to the river water supplying agency and approximately Rs 1 lakh per month for the water received from the local tank. The water maintenance charges are approximately Rs 33 lakhs per month. In all, approximately Rs 1,16 lakhs are spent for water by the local body per month in which overhead charges are not included.

On record the statistics of water taps in the town is not available. However, as per the number of authorized occupants the total taps could be 4000 including common water taps. A survey of the town reveal the fact that the actual taps are more than 15,000; the number of tubewells and pumps are also just double the official estimates. Besides, unauthorized quantity of water extracted from government tubewells, supply through water tankers is also a flourishing business in the town. Each tanker carries approximately 6,000 litres and on an average 200-250 trips are made by water tankers and tractor trolleys to supply water on payment. Taken together, the total consumption of water in the town comes to more than 80,00,000 litres per day. The excess quantity of water is supplied through unauthorized ground water supply ignoring legal and ecological restrictions on underground water exploitation.

Among the unauthorized occupants; 40 per cent belong to below poverty line and low middle class who have access to authorized public tapes and hand pumps, 30 per cent belong to lower middle class who largely depend on authorized taps for water supply and 20 per cent middle class and 10 per cent are among upper class who have constructed unauthorized multistorey buildings or bungalows, having access to both authorized wells as well as unauthorized water supplies through taps and/or tube-wells. The water supply to all inhabitants of the town is common and at the same pressure and for the same duration. However, middle and upper classes use illegal means such as placing water pumps on taps to extract more water from the common supply.

In other words, the maximum unauthorized use of water is by upper class who has means and ways including corrupt practices and money to get access to more than required quantity of water. A sample households survey in terms of water consumption reveals the fact that approximately, 40 per cent of the available water in the town is consumed by the upper class and about 35 per cent is consumed by the middle class. Though the middle and upper classes constitute 30 per cent of the total population of the town, they together consume 75 per cent of the available water while 70 per cent remaining population has access to only 25 per cent of the total water supply. This segment of the population does not have even the water storage capacity for more than 10 ltr per person per day.

In all the available daily water supply in the town is 80,00,000 litres per day. Accordingly the per capita availability is 120 litres. However, only 40 litres per capita water is accessibility to 70 per cent population of the town. In other words, 290 litres per capita water is accessible to middle and upper classes, which is about seven times more than the actual water accessible to lower classes. If the available water is equally distributed and become accessible to all classes than the per capita accessibility would be 120 litres which is sufficient to meet the basic need of water to all people.

It can, therefore, be concluded that the total water requirement of the town could be easily met from the existing supply. However, due to illegal means the population particularly economically poor sections suffer more from water shortage mainly due to unequal access to water supply. Since the middle and upper classes have influential relations with the local administration, the poorer sections suffer more from the water scarcity and unequal access.

Case 2: Industrial Water Use

Paper, distillery, tannery and textile are among the major water consumers. For presenting a case, a textile factory is selected. The textile factory 'Y' is located in the water scarcity region. However, being a sister concern of a big house in business and industry world and also due to comparatively much greater environmental concern, the factory looks like a green oasis in the desert area. The factory is selected to present as the case study because the effluent treatment plant and recycling and reuse of treated water in the factory could be projected as an ideal type of recycling of water resources.

Being a textile factory the water demand is comparatively higher. Besides, there are about 500 families residing within its premises. There are about 10,000 trees and more than 20 lawns and green belts along the roads with efficient irrigation systems. Almost all the families residing within the campus maintain lawns and kitchen gardens. The main source of water supply is a river located at a distance of about 7 kms from the factory. There are borewells also within the factory premise; however, number of borewells could not be ascertained.

The daily water requirement of the factory is about 30 lakh litres of which 90 per cent is received from the river through a pipeline maintained by the factory and remaining 10 per cent is met from borewells. Nearly 75 per cent of the water in the factory is used for industrial purpose mainly processing (80%) and cooling (20%) and 25 per cent is consumed for domestic purposes including gardening. The factory treats its effluent with efficient systems and claims to recycle 100 per cent treated water of which 60 per cent is used in gardening.

Water metres are provided at the initial supply points. The water cess is charged at the rate of Rs 9.5 per kilolitres for water used in processing which gets polluted with pollutants that are not easily biodegradable; Rs 7.5 per kilolitres for water used in processing which gets polluted with pollutants that are easily biodegradable; Rs 3 per kilolitres for domestic purposes and Rs 2.25 per kilolitres for water used in cooling and boilers. However, the water cess is paid for 2-3 lakh litres daily. Another important aspect is free access to river water without any check whatsoever. These two points need serious discussion.

Case 3: Groundwater

Dependency on groundwater is increasing at a tremendous rate. The levels of underground water have gone deeper and deeper leading to serious ecological consequences. At present, controlling and legislative measures to check underground water exploitation have become ineffective. In my opinion, underground water should be treated as ecological capital and should not to be used for human consumption because consequences of groundwater depletion are not properly known which may lead to any disaster unimagined. Nobody pays for the underground water. Ownership of surface land does not provide ownership of depth of the owned land including underneath resources. Land price is paid in square units for length and breadth and not in cube units to include depth also. There should be important changes in national water policy in terms of favouring environmental economics giving more monetary value to environmental conservation than commercial use.

Case 4: Drought

The largest portion of water use is in agriculture. There is a perception of water shortage, especially during drought periods. But this perception is largely the result of existing institutions for water management, institutions that were designed in the period of abundance. Drought is a normal, recurrent feature of climate, although many erroneously consider it a rare and random event. It occurs in virtually all climatic zones, but its characteristics vary significantly from one region to another.

Drought is considered largely a natural event and accordingly it becomes important to define the exposure (i.e., frequency of drought of various intensities and durations) of various parts regions to the drought hazard. A reliable assessment of water availability and its outlook for the near – and long-term is considered valuable information in both dry and wet periods. In country like ours, drought is more concerning with agriculture.

Agricultural drought links various characteristics of meteorological (or hydrological) drought to agricultural impacts, focusing on precipitation shortages, differences between actual and potential evapotranspiration, soil water deficits, reduced groundwater or reservoir levels, and so forth. When drought begins, the agricultural sector is usually the first to be affected because of its heavy dependence on stored soil water. Although drought is a natural hazard, society can reduce its vulnerability and therefore lessen the risks associated with drought episodes. The impacts of drought, like those of other natural hazards, can be reduced through mitigation and preparedness (risk management).

Planning ahead to mitigate drought gives decision makers the chance to relieve the most suffering at the least expense. Reacting to drought in "crisis mode" decreases self-reliance and increases dependence on government and donors. And this leads to more hardships. Some areas are considered more at risk than others. Vulnerability, on the other hand, is defined by social factors such as land use patterns, government policies, social behaviour, water use, population, economic development, diversity of economic base, cultural composition, and so forth.

It is argued, therefore, that we do not have so much a water scarcity crisis as we do a water management crisis. In terms of the water management during drought and its conservation for agriculture, industrial and domestic purposes, it is not the hydrological cycle (ecological) but the hydro-illogical cycle (socio-economic and political) which concerns more. Therefore, drought is an insidious hazard of nature having detrimental consequences on people more due to unsustainable practices, ineffective socio-economic structures and functions to cope up the hazard.

Therefore, drought should not be viewed as merely a physical phenomenon of water scarcity or natural hazard. Its impacts on society result from the interplay between a natural event (less precipitation than expected resulting from natural climatic variability) and the demand people place on available water supply.

Human beings often exacerbate the impact of drought more due to their own social fabric rather than nature. A guiding philosophy to ensure that its citizens will have dependable water resources and adequate water supply, conflicting policies can be reconciled, and the public interest can be better served, particularly during times of water shortage. Without a coherent guiding philosophy, the jumble of agencies assigned various responsibilities regarding water can actually work to inhibit water stewardship, drought planning, and routine water management decisions.

Another major deficiency of water resources planning in our country has been its fragmentation. Water planning has not been adequately integrated with land use planning; planning for river basins has neglected the water-related needs of metropolitan areas; environmental planning, or water quality planning, has been separated from water resources development planning; and agencies' plan have been project-specific, mission-specific, and uncoordinated with water resources plans of other agencies.

Long-term mitigation measures, such as implementing policies that require conjunctive use of ground and surface water, may require drafting new legislation and finding funds to support new monitoring and regulation efforts. In any case, it is essential to recognize that reducing long-term vulnerability to drought will require a sustained effort, although it may be a matter of long-term programs undertaken by a variety of agencies. Policy and planning for water conservation are faced with the following obstacles:

1. Politicians, policy makers, and the general public lack an integrated and holistic understanding of water resources and locally available and indigenous means for their conservation.
2. In areas where water resources are in depleted and polluted conditions, governments ignore integrated planning, or give it low priority or the peoples' representatives find themselves more concerned with the short term political interests instead of long term societal interests.
3. Inadequate financial resources are made available for a comprehensive implementation of conservation plans.
4. The policy resources and government machinery works on a stereotyped pattern in all areas.
5. Responsibilities are divided among many governmental jurisdictions.
6. We, and in that sense most countries lack a unified philosophy for managing natural resources, including water.

Social, economic, and environmental values often clash as competition for resources scarcity intensifies. Thus, scarcity of resources is more a social concern than a natural scarcity of resources. Water resources, their use and planning calls for integrated development strategies.

Sustainable Development

The concept of sustainable development was used for the first time in the World Conservation Strategy Report. The report was prepared by the WCED created by the UN as a result of a General Assembly resolution in autumn, 1983, and chaired by Mrs. Gro Harlem Brundtland, the then Prime Minister of Norway.

The World Commission on Environment and Development defined the concept as paths of human progress which meet the needs and aspirations of the present generation without compromising the ability of the future generations to meet their needs

Sustainable refers to a number of concepts like sufficiency, steady-state economics and, the notion of seven generations as the basis of decision-making. Sustainability to which we are concerned here entails meeting the need of the present generation without compromising the ability of future generations to do so. The simplest expression of sustainability involves asking in regard to each policy or decision, "Will our grandchildren and further generations also be able to enjoy living in the way we are living?"

In this regard many of our current practices are unsustainable because, chemical-based modern agriculture which depends upon limited resource of petroleum and which is rapidly eroding our top-soil is not sustainable; fossil-fuel based transport and energy system are unsustainable; the present economy that annually pours billions of tons of pollution into our environment making it unhealthy, is not sustainable; and the ever growing consumerism along with multiplying exploitative modes is not at all sustainable.

In other words when the modern way of living is analyzed nothing seems to be sustainable. The way society treats its children reflects its sense of justice, its commitment to the future and its urge to enhance the living conditions for coming generations. Environmental resources which we exploit for our development are also the basis for development of future generations.

However, the present generation has engaged itself in an ever accelerating indiscriminate exploitation of valuable natural resources without any consideration to the ecological consequences thereof and the needs of the future generation. The current trend in population growth, consumerism and socio-ecological disorder raises serious concern about the long term sustainability of the ecosystem. The consequences of environmental degradation are not the inevitable result of the technological progress only but rather the result of application technologies that are increasingly powerful but still crude, resource depleting, inefficient, wasteful and value ignoring.

The present ecological crisis and social disorganization pose serious problems about future. Research on role of images of the future generations suggests insights into our present social settings. If images of future do play an important role in social dynamics, and if our prevailing images are losing their dynamism because they lag behind development in society in general and of the weaker sections in particular with regard to developing economy and technologies, our own society is beginning to experience a crisis of direction.

Resources allocation should facilitate uses that provide for the greatest sustainable benefits and promote the transition to a sustainable and integrated management of resources. In doing so environmental, social and economic issues should be taken into consideration. Protected areas, private property rights, the rights of the indigenous people and local communities and the economic role of women in agriculture and rural development among other issues, should be taken into account.

If in the future, human requirements are to be met in a sustainable manner, it is essential to resolve these conflicts now and move towards more effective and efficient use of land and its natural resources. Integrated physical, social and land use planning and management is an eminently practical way to achieve such goal. Integration should take place at two levels, considering all social and economic factors on the one hand and all environmental and resource components together, on the other but simultaneously.

Thus, sustainable development should be taken to mean a thriving socio-economic order wherein the production structures and relationships ensure fair distribution of resources and opportunities based on the principles of equality, liberty, justice and fraternity along with the long-term carrying capacity of the ecosystem with minimum negative impacts on the society. Therefore, until and unless a congenial social environment is created, the fruits of development cannot be sustained.

From the foregoing discussion it becomes clear that the concept of sustainable development has to be consistent with justice to nature; justice in respect of socially, economically and ecologically disadvantaged sections of our society; and justice between gender generations, i.e., intra and inter-generational.

Sustainability is, therefore a much broader phenomena enhancing ethical norms pertaining to the survival of living beings, to the right of the future generations and to institutions responsible for ensuring that such rights are fully taken into account in policies and action.

In the ultimate analysis it can be said that if we are to meet the needs of both the present and the future generations, substantial shift in values, attitudes and behaviour must occur. The most important change in the global order after the World War II was the shift from a primarily political world system to a primarily economic order. Now again we are experiencing a transition in the worldview towards the ecological balance. At present, there is little discussion about alternatives.

Environmental thinking has to be different in the changing scenario, from a dominant mode of thought in modern industrial society. The scientific and technological component of modern society has become so phenomenally successful that people in most societies expect more from technology than it can deliver. Transforming society to a sustainable state, however, is not just a technological matter. It will require wide and deep social learning of thinking, value structure, behaviour pattern and institutional arrangements. People must learn such basic like systematic thinking, the two laws of thermodynamics, bio-diversity for ecosystem stability, carrying capacity, environmental sources and sinks, the tragedy of commons, the distinction between growth and development and inter-generational equity.

Constitutional Status of Water Resources

Constitutionally, most of the authority for allocating water resides with the government which is also the biggest player in natural resource planning, management, regulation, and development. Barring ground water extraction from private land, allocation of surface and ground water resources like any other natural resource, is under discretion of government, central or state and/or its instrumentalities.

Throughout the world, authorities responsible for allocating water make their decisions based on a number of doctrines. Surface water distribution is based on either the riparian doctrine or the prior appropriation doctrine. The riparian doctrine holds that owners of land adjacent to a stream or other water body may use that water, but only as long as their usage do not interfere with the other landowners' reasonable usage of the water. In contrast, the prior appropriation doctrine gives rights to the earliest water users over all later water users of a particular source – "first in time, first in right."

For groundwater, water allocation is determined by three doctrines: absolute ownership, the reasonable use doctrine, or the appropriation-permit system. Absolute ownership gives rights to the overlying owner and does not restrict usage. The reasonable use doctrine gives groundwater rights to landowners, but with the added provision that their use of the water cannot interfere with other users. The third doctrine, the appropriation-permit system, holds that groundwater rights are based on priority, meaning that prior users of groundwater have the greatest legal rights

During industrialization and specifically after new economic policy governments, irrespective of their political ideologies encouraged development in part by ensuring that the new landowners have access to as much resources as they needed (the doctrine of prior appropriation). This is a part of a distinctive body of policy and law based on an extraordinary combination of two ideas: that public resources should be made available for private gain free or at far below market value; and that the government, in addition to these initial subsidies, should further fuel the development by affirmatively building water projects and other public works to support the opening of the privatization

On comparison with land, as long as settlers used land, they could own it. However, once all the land was allocated and it became scarce, the nation made the transition to a market. After transformation of land from a natural resource to economic resource, today we are in the same situation with water resources as well. It is argued that a more market-based system for allocating water would provide the necessary incentives for conservation in an era where economics and scarcity conditions will play a greater role in land and water resource management.

Our Constitution has mixed provisions for waterways and water resources. Under the Seventh Schedule (Article 246) of our Constitution, water is taken care of under entries 56 and 97 of Union List and 14, and 18 of State List. Entry 56 of List I is wide enough to include the use, distribution and allocation of the waters of inter-state rivers. Under entries 14, and 17, of List II, the state legislation can legislate with respect to all waters within its territory, but so far as inter-state rivers flowing through the state is concerned, the power of the state legislation is subjected to entry 56 of List I. Article 262 of the Constitution specifically deals with the disputes relating to waters. The article specifies that:

1. Parliament may by law provide for the adjudication of any dispute or complaint with respect to the use, distribution or control of waters of, or in, any inter-state river or river valley.
2. Notwithstanding anything in this Constitution, Parliament may by law provide that neither the Supreme Court nor any other court shall exercise jurisdiction in respect of any such dispute or complaint as is referred to in clause (1). In exercise of the power conferred by this article, the Parliament has enacted the Inter-State Water Disputes Act, 1956.

Section 4 (1) of this Act provides that when, on receipt of a request from any State Government, the Central Government is of opinion that the water dispute cannot be settled by negotiations, The Central Government shall constitute a Water Dispute Tribunal for the adjudication of that dispute.

In other words, the constitutional position on water dispute is that the State Legislature's powers relating to regulation and development of inter-state waters shall be exercisable only so long so far the control over their regulation and development has not been taken over by a law of Parliament passed under the entry 56 of List I. An altogether new thinking shall be required to review the Constitution with regard to water.

Our past experiences show that existing constitutional provisions are not sufficient and efficient to handle inter-state and centre-state issues like waters. Indiscriminate exploitation of ground water has resulted in lowering of water table. In the coming future, it will be difficult to cope with the water problems and it is also said that the next world war will be fought on water!

Dr Ambedkar's Policy for Conservation and Management of Water Resources

The recent Cauvery river water issue has resulted into unprecedented tension between the states of Karnataka and Tamil Nadu. Continuing problem over Narmada river between the states of Gujarat and Madhya Pradesh appears to be different from Cauvery issue but has the same roots. Similarly, recurring floods and droughts in our country are considered as natural disasters but it is time now to think how and why these are not natural disasters but are man made.

Interestingly, Dr Ambedkar whose personality is usually limited to only a *Massiaha of Dalits* or the Father of the Indian Constitution by both Ambedkarites and non-Ambedkarites alike, was a great visionary of the betterment of humanity and sustainable development. He was very much aware of the importance of natural resources for the sustainable development of society in a much better way than any other social scientist or environmentalist having an understanding of current problems.

The period from early 1920s to mid-1950s was of crucial importance due to freedom struggle; the process of Constitution making and economic and social framework of modern India was initiated. Ambedkar addressed and contributed practically on all the issues of national and foreign importance that arose during that period in capacities and as a leader of depressed classes which were the concerned target for policy framework. The Department of Industry and Labour was bifurcated to create an independent departments of Industry (including Commerce) and Labour (including irrigation, electricity and other works related to public works) in November, 1937.

Ambedkar took over the Labour portfolio in July 1942. (NAI, 1987). It was the time when the planning for the irrigation, waterways and related issues began. During 1942-47, a policy for water resources and hydro-electric power development was evolved, an administrative infrastructure was created and high level technical organizations like Central Waterways, Irrigation and Navigation Commission (currently known as the Central

Water Commission) came into existence. In fact, the present day policy and planning for multipurpose large dams is a continuity of the initiation on conceptualizing water resources for national use during the period referred to above.

Ambedkar being at the helm of affairs of departments concerned with the water resources was instrumental and made certain fundamental contributions in evolving a water policy for our country. Thorat (1998) draws a comprehensive picture of the various alternative plans and committees for the reconstruction and economic development during 1940s and 1960s and discusses the origin of water policy for the development of irrigation, waterways and drainage; development of electric power; and inland water transport.

The policy statement framed by the Labour Department under the leadership of B.R. Ambedkar advocated for a greater role of the Centre in developing water resources as against the power given to the Provinces under 1935 Act. Dr Ambedkar knew it better and had a clear vision of such policy. He considered natural resources as common human property with no provincial barriers. In fact, environmental resources observe no political boundaries. And if political boundaries are artificially made for them then problems arise. Dr Ambedkar had a very clear perception about the community or joint-ownership of water resources across the land. He considered them as property to be shared by all to the best advantage.

His vision becomes clear from his address to the representatives of Centre, Bengal and Bihar Governments in Bengal Secretariat in Calcutta on 3rd January, 1945 (*Dr Babasaheb Ambedkar: Writings and Speeches*, Vol. 10, 1992: Dr Ambedkar as member of the Governor General's Executive Council, 1942-46, Education Department, Government of Maharashtra, p 220):

> The Government of India is very much alive to the disadvantages from the present state of affairs and wishes to take steps to evolve a policy which will utilize the water resources of the country to the best advantage of everybody and to put our water resources to the purpose which they are made to serve in other countries.

Not only he was concerned with the water resources for national development, Dr Ambedkar explicitly and implicitly emphasized that water resources were the concern of Central Government and not that of Provincial Governments.

He made particular reference to one of the recommendations of Damodar River Flood Control Committee appointed by the Government of Bengal in 1944:

> During the course of discussion it was felt by the committee that it will be an advantage in the solution of flood control and soil conservation problems if forests and rivers of India are made the concern of the Central Government

Dr Ambedkar considered that since water resources observe no political boundaries and were often found in more than one province, they could not be subjected to provincial boundaries. The present water wars over Cauvery and Narmada rivers are in fact the result of such policy only where rivers are kept in both the Union and State lists. At the Damodar Valley Scheme Conference at Calcutta in 1945, Dr Ambedkar said:

> [W]e have not taken sufficient account of the fact that there is no difference between railways and waterways and if railways cannot be subjected to provincial boundaries neither can waterways at any rate those that flow from province to province. The disadvantages of this error are many and obvious.
>
> To give one illustration, a province needs electricity and wishes to utilize its water resources for the purpose but it cannot do so because the point at which water can be dammed lies in another province which being agricultural does not need electricity and has no interest in it, or money to finance the project, and would not allow the needy province to use the site. Complain as much as we like, a province can take such an unfriendly attitude and justify in the name of provincial autonomy.

Thus, Dr Ambedkar had a clear perception of geographical inequalities of natural resources and recommended a centralized approach to minimize such inequalities. Hence, for securing environmental justice to society in general and weaker sections in particular, there is an urgent and immediate need that not only water but all basic life sustaining resources in the country should be distributed equally and to the best advantage of all, as envisaged by Dr Ambedkar in his policy and vision on water.

The unity in diversity of our country will be strengthened if a uniform policy for life sustaining basic resources is adopted. Recurring floods and droughts have become unmanageable and can be tackled by adopting a uniform, integrated and coordinated policy, which can only be designed, implemented and ensured by the Central Government. Dr Ambedkar was of the opinion that

> [I]t is wrong to think water in excessive quantity is an evil. Water can never be so excessive as to be evil. Man suffers more from lack of water than from excess of it. The trouble is that nature is not only niggardly in amount of water it gives, it is also erratic in its distribution alternating between drought and storm.
>
> But this cannot alter the fact that water is wealth. Water being wealth of the people and its distribution being uncertain, the correct approach is not to complain against the nature but to conserve the water.

Dr Ambedkar considered water as property to be shared by all to the best advantage. His policy to treat water particularly inter-state water ways (rivers) as a matter of concern for the Centre like railways and not that of province, is the solution to current inter-state conflict over waters like that between Karnataka and Tamil Nadu and Madhya Pradesh and Gujarat.

All basic life sustaining natural resources should be declared and treated as national resources ensuring equal distribution and accessibility to all the citizens a prerequisite for socio-ecological democracy. There should be only one policy (national policy) for water and in that regard for all life sustaining resources and also there should be only one agency (central with due representation from all states and union territories) to implement, regulate and monitor the distribution and accessibility to such resources to the best advantage of all.

Interlinking of Rivers

Mega Mother-project Planning of Water Resource Conservation in India

Ecologically, and on a large scale, water conservation depends on the health of watersheds, which are areas bounded by the divide line of water flows so that a distinct drainage basin of a water course can be identified. The rain falling over this area will flow through only one point of exit of the whole watershed.

There are mainly three watersheds in our country viz. Himalayan range with its Karakoram branch in the north, Vindhya and Satpura ranges in Central India and Sahyadri or Western Ghats on the west coast. All the major rivers of India originate in one or the other of these watersheds.

Recently, the riverlinking project has been planned by the Government of India which is expected to be completed by 2016. The project will connect the rivers in the north with those of the south through a network of canals. There would be 30 links having 14 Himalayan and 16 peninsular components. The task force constituted for the purpose, plans to attract volunteers who would be given green cards in villages to help in water conservation. On the basis of 30 feasibility studies conducted, the National Water Development Agency indicates that the project is viable, since the canals would be largely based on gravity with greater storage facilities.

Water from the Brahamputra would flow into Ganges which in turn would be connected to Mahanadi and Godavari which would be linked to Krishna, then to Pennar and Cauvery. Similarly Narmada would flow into the Tapti and Yamuna into the Sabarmati. The project would benefit both the flood-prone as well as drought hit areas of the country. New storage facilities besides ensuring even distribution and availability of water, would provide food security by increased irrigated area and also additional 34,000 KW electricity. At present, 24,000 KW electricity is being generated.

In other words, it is a very big project and if commissioned, would certainly solve most of the present problems related with water and land. Even if hiccups of environmentalists in any form are taken due care, this very big project and viable strategy of water conservation needs a reorientation in terms of constitutional status of water, values of consumerism and social participation. The project can change the scenario of land and water resources of our country provided ecological sustainability is given priority over social, political and economic gains.

In conclusion, against the prevalent processes of decentralization, in the interests of the weaker sections and sustainable development, there is an urgent need to socio-politically and ecologically centralize all the natural resources so that a uniform policy and planning could be evolved for the conservation and management of our natural resources including land and water giving more consideration to environmental economics and national integration. In other words all natural resources should get entry into Union List of the Constitution alongwith changes in the policy towards natural resources conservation.

The national water policy should give importance to environmental economics, giving more value to environmental conservation than to water's commercial use. The price of underground water should not be in terms of two dimensional purchase of land (length and breadth) but it should be three dimensional including the price for depth also. This will serve two purpose, firstly making the underground water much costlier and beyond the reach of almost 90 per cent of the population, and secondly, even then the affordable class (10%) prefer to purchase underground water, the money so collected could be utilized for the conservation of water resources. Because the underground water is to be conserved at all cost for the sustainable development.

■ References

Adiseshiah, Malcolm S. (1990): 'Sustainable Development: What It Is and Its Coverage'. In *Sustainable Development: Its Content, Scope and Prices,* Lancer International in association with India International Centre, New Delhi.

Attfield, Robin (1983): *The Ethics of Environmental Concerns*, Columbia University Press, New York.

Coleman, Daniel A. (1994): *Ecopolitics: Building a Green Society*, Rutgers University Press, New Jersey.

Das, Bhagwan (1979): *Thus Spoke Ambedkar.* Vol. 3. Bheem Patrika Pub., Jalandhar

Dr Babasaheb Ambedkar: Writings and Speeches, Vol. 10 (1992): Dr Ambedkar as member of the Governor General's Executive Council. 1942-46, Education Department, Govt. of Maharashtra, pp. 220-221, 305.

Dworkin, R. (1981): "What is Equality? Part I: Equality of Welfare and Part II: Equality of Resources", *Philosophy of Public Affairs.* 10 185-246 & 283 – 345.

Hargrove, C. Eugene (1989): *Foundations of Environmental Ethics*. Prentice Hall, New Jersey.

Herman, Daly (1983): "On Sustainable Development and National Accounts," in *Economics, Growth and Sustainable Environments*. (Eds. David Collard, David Pearce and David Ulph), St. Martin's Press, New York.

Hughes, D. (1983): "Gaia: An Ancient View of Our Planet", *The Ecologist*, 13 (2, 3).

Khoshoo, T.N. (1986): Environmental Priorities in India and Sustainable Development. Presidential address, Indian Science Congress Association, 73rd session.

Lashof, D. (1986): *The Ecology of Socialism*. Energy and Resource Group, University of California.

Leekok, Lee (1989): *Social Philosophy and Ecological Scarcity*. Routledge, London.

Naess, A. (1983): "The Shallow and The Deep, Long Range Ecological Movement", *Inquiry*, 16.

Norton, B. (1984): "Environmental Economics and Weak Anthropocentrism", *Environmental Ethics*, 6(2).

Olson, Robert L. (1 995,): "Sustainability as a Social Vision", *Journal of Social Issues*, Vol. 15, No. 4, 15-36.

Passmore, John (1980): *Man's Responsibility for Nature: Ecological Problems and Western Traditions*, London

Pearce, David W. and R. Turner (1990): Ethics and Future Generation, pp. 209-23 In: *Economics of Natural Resources and the Environment*, Harvester Wheatsheaf, New York.

Pepper D. (1984): *The Roots of Modern Environmentalism*, Croom Helm, London.

Rawls, J. (1971): *A Theory of Justice*, Harvard University Press.

Sandbrook Richard (1992): "From Stockholm to Rio". In *Earth Summit, 1992*, (Ed. Joyce Quarrie), The Regency Press, London.

Smith, David (1994): *Geography and Social Justice*. Blackwell Publishers, Oxford.

Sunderam K.V. (1992): "District Planning for Sustainable Development". *Annals NAGI* Vol. XII No. 1 & 2

Sylvan, R. (1985): "Deep Ecology", *Radical Philosophy*, 40/41.

Thorat, Sukhadeo (1998): *Ambedkar's Role in Economic Planning and Water Policy*. Shipra Publications, Delhi.

Timothy, O'Riordan (1985) "Future Directions in Environmental Policy", *Environmental and Planning*, 9(3).

Timothy, O'Riordan (1988): "The Politics of Sustainability", In *Sustainable Environmental Management*, Belhaven Press, London.

UNDP Human Development Reports, 1990-1998: Oxford University Press, Oxford.

United Nations (1982): Ten Year Report (1972-82), United Nations Environment Programme.

Verma, D.K. (1994): "Ecoethical Approach for Sustainable Development and Justice", *AJSDJ* Vol. 2., p 74-80.

Verma, D.K. (1995): "Ecological Sustainability in Social Development". In *Social Development and Justice,* Eds R.G. Singh and R.D. Gadkar, Radiant Publishers, New Delhi.

Verma, D.K. (1997): "Dr Ambedkar's Views on Planning, Development and Conservation of Environmental Resources". *AJSDJ,* Vol. 5, 1-8.

Verma, D.K. (1999): "Concept and Dimensions of Social Environment Needed for Sustainable Development". In: *Social Environment and Sustainable Development.* Ed. Y.G. Joshi and D.K. Verma, Rawat Publication Jaipur.

Verma, D.K. (2000) *Sustainable Development.* Vol. 4 (2), pp. 37-59, July-December, 1999.

Verma, D.K. (2002): *Dr B.R. Ambedkar Ke Prakratik Sansadhano Ke Dohan Prabhandhan Yojna Aur Paryavaran Sanrankshan Par Vichar*, 8-28 April, 2002, p. 7 (2nd Edition).

Verma, D.K. (2002): "Dr Ambedkar's Policy: Solution to Cauvery Water" Dispute" (Letter to Editor, Free Press, December 13, 2002 columns 1-3, p. 6).

World Commission for Environment and Development Report (1987): *World Conservation Strategy Report*, 1987.

Chapter 15

Interlinking of Rivers

Prospects and Problems

Raj Kumar Siwach

THE Kalinga incident in Orissa, the Polavaram project in Andhra Pradesh and the Narmada dam controversy have recapitulated intense discussion over the fate of such projects in the context of environmental sustainability, social approval, political acceptability and financial viability. The major purpose of these projects is to enhance the socio-economic status of the people.

The Inter-Linking of Rivers (ILR) project is one of such projects, launched by the Government of India, to contain twin problems of droughts and floods by way of transferring surplus water to deficient regions of the country through 30 major river links over 12,500 km of canal length and construction of 32 dams, equally, needs a thorough review of scientific rigorousness versus environment debate in terms of cost-benefit analysis. So, it becomes imperative to comprehensively look into the pros and cons of the project.

The project envisages the implementation of the National Perspective Plan under the supervision of the National Water Development Agency (NWDA). For giving this plan a concrete shape, the NWDA, so far, has completed all the pre-feasibility studies, including alternative study of basin hydrology of all 30 links. The feasibility studies for 14 links in the Peninsular component and two links in the Himalayan component have also been completed. On the basis of these studies, the ILR project, *prima facie,* is technically feasible and economically viable. The work on political consensus building is going on for Ken-Betwa and Parbati-Kalisindh-Chambal links involving Rajasthan, M.P. and U.P. Thereafter, the Detailed Project Report (DPR) will be finalized, which shall incorporate, among other things, financial analysis, environmental management plan, rehabilitation and resettlement package, social impact analysis of the affected persons, engineering assessment, comprehensive data base, governance structure for implementation and performance review, time schedule and final report.

Thus, the project needs a scientific enquiry with a view to investigate its environmental and development impact over the people, plants, demography and the eco-system. Thus, an attempt has been made in this paper to analyze the pros and cons of the ILR project.

Relevance of the ILR Project in the Context of Water Scenario in India

The total water availability in country has been calculated as 4,000 billion cubic metres (bcm), of which 1,869 bcm is available as run-off. Of this, 1,222 bcm is utilizable. However, till now, storage capacity for only 177 bcm has been developed, 208 bcm being in the pipeline. This situation will exacerbate, if no preventive measures are taken. In future, water demands for irrigation, drinking, industries, energy and others will be 813 bcm, 1,093 bcm and 1,447 bcm in 2010, 2025 and 2050 respectively. The projected water use, per capita, per year, in year 2050, would be between 725-750 m^3, as compared to about 650 m^3 at present.

Further, the rainfall variations pose a serious problem, as some parts of the country, e.g., the states of Assam, Bihar, West Bengal, Orissa and the U.P. receive much more rainfall causing heavy floods, whereas the states of Rajasthan, Gujarat, M.P., Haryana, Tamil Nadu and Andhra Pradesh witness normal, sometimes less than normal, rainfall causing droughts. Thus, while one region of the country is inundated, the other clamours for water.

Indian river basins also show wide variations of run-off. For instance, over 90 per cent of annual rainfall in the Peninsular rivers and over 80 per cent of the Himalayan rivers occur during the four monsoon months. For this temporal variation of run-off, most of the water not only flows unutilized to the sea but also causes immense flood losses.

Keeping in view this emerging picture of water resources, the idea of interlinking of rivers for transferring surplus water to deficient areas seems to be an innovative strategy for containing floods, droughts and accelerating the speed of socio-economic development initiatives.

Hydro-Power Generation

The Himalayan component, if implemented, would provide additional irrigation to about 22 million hectares and generation of about 30 million KW of hydro-power, besides providing substantial flood control in the Ganga and Brahmaputra basins. It would also provide the necessary discharge for augmentation of flows at Farakka, *inter alia*, required to flush the Kolkata Port and the inland navigation facilities across the country.

The Peninsular component envisages construction of as many optimal storages as possible on these streams and interlinking them to make available appreciable quantum of water for transfer to areas where additional water is needed. The scheme provides for taking water supply canal to the metropolitan areas of Bombay and irrigation to the coastal areas in Maharashtra.

The prospective benefits of this component would generate additional irrigation to about 13 million hectares and generation of about 4 million KW of power.

Enhancing Foodgrains by Irrigating Large Tracts of Land

Due to irrigation by river water and ground water, the country has achieved the target of 200 million tonnes of foodgrains at present. The irrigation area has been increased to 95 million hectares. The present population is expected to increase 1,500 to 1,800 million in the year 2050, which will require 450 million tonnes of foodgrains. For achieving this target, it is necessary to increase irrigation potential to 160 million hectares for all crops by 2050. But, our maximum irrigation potential through conventional sources has been assessed only about 140 million hectares. So, we should evolve new strategies to attain the potential of irrigating 160 million hectares of land for all crops by 2050. The proposed project, therefore, is considered as an imperative strategy to meet the requirements of irrigating large tracts of the land.

Earlier Successful Examples of Irrigation Projects

The rationale of the ILR project could also be justified as some other countries also have followed the same suit. These are: the China's Three George Project across the Yangtse to be completed in 2009 involving Rs 1,50,000 crore outlays and the State Water Project for transferring water of Sacramento river and the Central Valley Project of California, USA, for supplying irrigation water to farmers by a vast system of dams, pipes and channels making the state's agriculture status as one of the most productive in the world.

Such hydraulic engineering projects in India also have ameliorated the socio-economic conditions of the people in an exemplary way by eliminating the droughts, poverty, squalor and meeting the needs of water, domestic power and industries in arid and semi-arid regions. These include the Indira Gandhi Nahar Pariyojna, the Periyar Project, the Parambikulam Aliyar Project, the Kurnool-Cuddapah Canal Project and the Telugu-Ganga Project.

Financial Viability

According to a tentative assessment of NWDA, the project's total cost is put as Rs 5,60,000 crore over a period of 12-15 years. It has three parts: Rs 1,06,000 crore for the Peninsular component, Rs 1,85,000 crore for the Himalayan component and Rs 2,69,000 crore for hydro-electric component. However, this amount does not take into account the cost of acquisition of land, cost of resettlement of displaced persons, capital cost for water lifting, interest and inflation calculation. If we assume 7 per cent rate of interest and 5 per cent annual rate in inflation, the total capital cost of the three components, at the end of 20 years, will be: Rs 3,81,878 crore for the Peninsular component, Rs 6,66,485 crore for the Himalayan component and Rs 9,69,105 crore for the hydro-electric component, thus totalling Rs 20,17,468 crore, approximately four times of tentative cost. It raises a question as to how and from where the huge funds will be mobilized. It has been suggested that Rs 4,65,000 crore could be raised by the ways of participation of public sector, private sector and public-private partnership in hydrological projects, irrigation infrastructure and annuity model in irrigation. Funds from financial sector will also be mobilized by issuing Green Bonds, retails investment, amnesty schemes, cess and duties on select goods of mass consumption, allocation of employment guarantee schemes and users fees from the benefited states.

The successful examples of raising resources in the project of Golden Quadrilateral of NHAI by toll tax, and other options of cess on petrol, surcharge on income tax by the Centre and sales tax by the states could make whole project financially viable and bankable.

Mitigating Drought Conditions

There are many drought-prone and water-deficient areas in the country. These are: Rajasthan, Gujarat, Maharashtra, Karnataka, Andhra Pradesh and Tamil Nadu. The frequent drought occurred in these states cause much misery to human beings, livestock, flora and fauna. It is estimated that 263 million persons, which are more than 26 per cent of the entire population, and 108 million hectares of land got affected by the recurring droughts.

This problem of droughts and water deficiency, if remains unabated, will unleash threats to inhabitants of other areas also. For instance, due to water deficiency compounded by cavalier use of domestic pumps in south and south-west Delhi regions, the Delhites will face acute water crisis in the year 2021, when 23 million people will need 1,800 MGD water against Delhi Jal Board's maximum augmentation capacity of 919 MGD.

Similarly, the ground water table in 286 districts across 18 states, according to a study, is being dangerously depleted. So, the future of water resources is beset with more conflicts and wars between communities, states and nations.

The ILR project, it is expected, will mitigate drought conditions by offering water to an area of 25 lakh hectares in affected states.

Containing Floods

In addition to the drought conditions, the country also witnesses the frequent occurrence of floods. According to an estimate, over 40 mha of the area experiences periodic floods claiming on an average 1,529 human lives and 94,000 cattles and loss of more than Rs 1,376 crore. Besides, the flood created havocs are: epidemics, hunger, starvation, migration, damages to forests, crops, plants, aquatic and wild birds and destruction of national heritage. The worst affected states include Assam, Arunachal Pradesh, Meghalaya, Uttranchal, Uttar Pradesh, Bihar, Orissa and West Bengal.

This natural calamity, it is hoped, will be contained under the ILR project, which foresees the construction of storage dams for regulating the inundation. The flood peaks, under the project, are estimated to reduce, not to eliminate altogether, by about 30 per cent in Ganga and Brahamaputra basins.

Boosting Industrial Growth

The economic benefits of the ILR project, both during implementation and post-implementation phases, in tangible and intangible terms, need not to be overemphasized. It is expected to boost industrial growth for civil construction and capital goods industry. It would enhance foodgrains production of about 8.5 crore per annum, 34,000 MW hydro-power capacity and fuel saving worth Rs 3,000 crore. Besides, spin-off benefits for agri-based industries, fisheries development, tourism promotion and plantation across the canal banks could also be visualized.

Health Savings

It has also significant social impacts over the lives of the people by 80 per cent increase in their household income by assured irrigation and saving of Rs 4,024 crore by the ways of lesser water-borne diseases, minimizing time for water-carrier housewives in rural areas and saving by the employers on health reimbursement. The direct annual saving of Rs, 4,000 crore is also anticipated. The National Council of Applied Economic Research (NCAER), in this context, has been organizing the studies for evaluating the project's impacts over national economy.

Employment Opportunities

The ILR project would generate employment opportunities for people in both rural and urban areas by engaging them in construction activities of dams, barrages, long canals, cross-drainage structures and power houses. Equally, the suppliers of cement, lime, stone and bricks and semi-IT literals in data collection and analysis, financial analysts and navigators will be benefited from the project.

Emotional Benefits

This project is considered as pragmatic not only for economic, social and financial transformations but also for strengthening emotional bonds of the people. The sharing of water is the symbol of the national integration. It will be, in the words of our Hon'ble President, "a key driver of the growth of the country not only for bringing economic prosperity but emotional integration as well." This will also act as a harbinger of resolving inter-basin conflicts amicably.

Inherent Challenges

No doubt, this national project has promising future for the prosperity and growth in the agriculture, irrigation, power and social upliftment. But, there are some inherent challenges confronting its workability. These are discussed below.

Civil Society-State Conflicts

The project has potential for generating conflicts between the state and civil society and local communities on the grounds of compensation packages for rehabilitations and environmental damages, riparian rights, sharing mutual benefits and costs. Their disputes will also become intense over compensation for loss of water, effects on existing power and irrigation generation, power needed to lift waters, acquisition of land, resettlement, resources loss, rational calculation for water lost *en route*, adverse impact on sediment load, silt and aquatic biodiversity.

Persuading Contending States: A Tough Task

One of the most challenging tasks, keeping in view the present political formations and compulsions, before the Union Government, therefore, is to build consensus among states, as water-surplus states will fight each other with tooth and nail. These contendings states like Kerala, Bihar, West Bengal, Assam Punjab, Chhatisgarh and Goa, have opposed the project while states such as Gujarat, Karnataka, Andhra Pradesh, Orissa, Maharashtra, Haryana and Tamil Nadu have supported the project conditionally. Even the legislatures of Kerala and Punjab have passed resolutions on inter-state disputes paving way for constitutional crisis.

However, more recently, agreements between Rajasthan, Madhya Pradesh and Uttar Pradesh over 24th and 25th links of the Peninsular component are good omen, But still, it is difficult to convince all the states for mutually sharing water. The states perceive 'water' as trump-card manoeuvrability for competitive state politics. The political class of the states views the electoral politics more important than nation-building efforts. The intense disputes between Punjab, Haryana, Tamil Nadu, Rajasthan and Karnataka are still pending.

Ecological Repercussions

The project, no doubt, has positive possibilities for, socio-economic change and development but its repercussions for the environmental degradation, including forests and land submergence, seismic hazards, topsoil loss, drainage congestion, spread of diseases, impact over flora and fauna, fisheries and wildlife, river morphology, climatological changes, decay of organic matter, cropping pattern, and aquatic and riparian life, have thrown a gauntlet before the policy makers and social scientists.

Resettlement Bottlenecks

In addition to environmental effects the rehabilitation, relocation and resettlement of the affected persons and consequences for their assimilation to new system, culture and beliefs are impediments in the way of social approval of the project.

The bureaucratic sloth, inefficiency and corruption in delivery of compensation packages, in the past, has created problems for 4,98,241 families affected by 16 Peninsular links.

Hence, it is absolutely essential to educate, enlighten and aware the citizens about all ecological and resettlement aspects of the project. We cannot pull the wool over public's eyes.

Legal Issues

In the smooth transfer of inter-basin water and integrated development of inter-state rivers, there are certain legal issues which should be addressed forthwith. The Parliament should pass an Act to include 'water', which is now in state subject, into Union List. The River Boards Act, 1956 provides only for Advisory Boards not for River Basin Organizations (RBOs) vested with powers of management. Thus, the Inter-State Rivers and River Valley (Integrated and Participatory Management) Act is strongly urged. This Act would provide for constitution of an RBO for each major basin. This organization should take initiative to bring the states concerned on a single platform for holding serious discussions on sharing of waters including diversion to non-basin states, paving way for reaching an agreement on the basis of mutual needs.

It shall also be duty of the RBO to collect data and disseminate them in local languages, formulate integrate plans and consider the proposals from constituent states on various issues including project proposals in the basin and monitor implementation of large projects.

The Central Water Commission (CWC) should be restructured into a statutory high-powered interdisciplinary agency with maximum autonomy, in order to deal with policy and reforms, centre-state and interstate issues, planning and project finalization, water conservation, environmental aspects and rehabilitation, people's participation and communication, facilitation of research, HRD, training and a National Information Data System.

There are a few Acts and rules which also need to be amended. These are Water (Prevention and Control of Pollution) Act, Environment Protection Act, 1986, and rules such as the Hazardous Wastes (Management and Handling) Rules.

Knowledge Gap

Due to lack of professional and scientific information on the entire ILR project, particularly the Himalayan rivers, there is need for more detailed study using systems analysis techniques. The Himalayan component's actual implementation for information confidentiality is unlikely to be undertaken in the coming years.

This knowledge gap, to some extent, could be bridged by adoption and application of innovations and research in the emerging fields of hydrometeorology, snow and lake hydrology, river morphology, and hydraulics, water harvesting, evaporation, seepage, tunneling technologies, advanced numerical analysis, seismic designs and construction technologies. More research efforts are also needed towards socio-economic aspects of water resources development emphasizing equity, employment and growth.

All research and development institutions should work together in a well-coordinated manner for more reliable and authentic findings, results and interpretations. There is no shortage of figures on water but the problem is how to interpret them

Tapping Traditional Wisdom

Not only technology diffusion but traditional wisdom of indigenous know-how about rainwater harvesting and conservation can yield rich dividends. Such community-based initiatives, in the long run, will enhance local augmentation of water availability. The shining examples of Rajendra Singh, Megasaysay awardee, in Alwar district in Rajasthan, Anna Hazare in Ralegan Siddhi, some NGOs in Gujarat, Dhan Foundation in southern states, as many as 39,000 tanks in Tamil Nadu and Karnataka, have all shown that local people can be good custodians of the nature. These people-centred movements could be more successful if the irrigation bureaucracy does not resort to archaic official hitches as experienced in *johads* in Rajasthan, and the local community is vested with rights of managing common resources.

Emerging Challenges for Water Administrators

The ILR project, particularly its operation, maintenance and implementation processes, have raised pertinent issues for the theory, practice and profession of Public Administration. The scarcity of the water and its demand side management have been slowly developing new paradigm for 'Water Governance'. In future, Public Programme Administrators, even more competent, will have to face daunting challenges of managing scare resources for meeting legitimate demands of ever rising population. Therefore, it requires extraordinary patience, skills and conviction of governing for offering solutions to the complex water governance problems stemming from rapid technological and demographic changes, holistic approach of eco-hydrology,

Figure 15.1

Proposed Inter-Basin Water Transfer Links

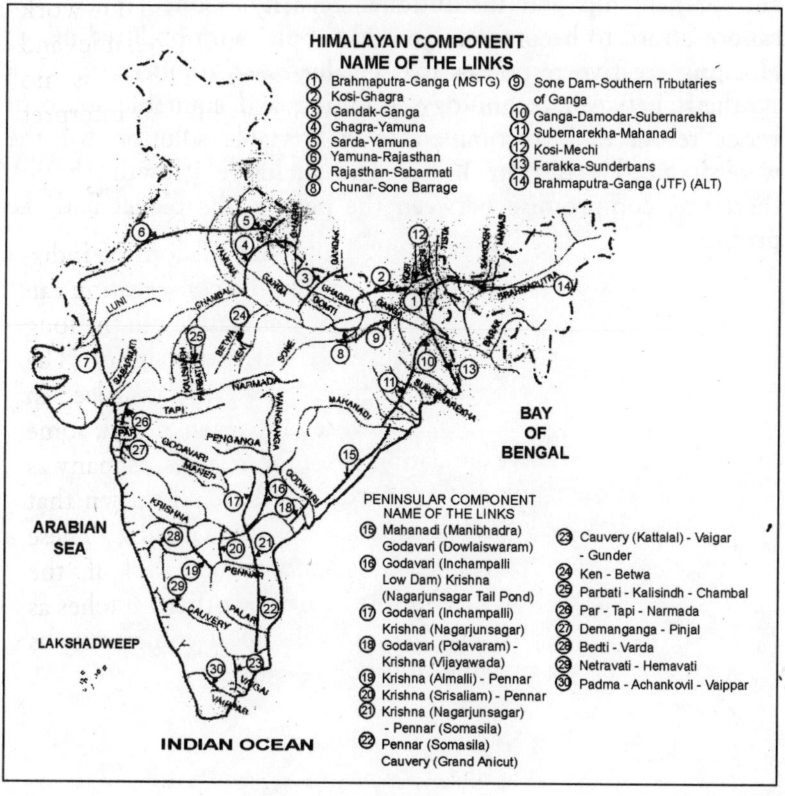

transdisciplinary knowledge, water harvesting community initiatives, water pricing and development-sustainability incompatibility. Thus, the prospective 'water administrators' will have to work hard to administer compensatory packages for fair distribution of water resources through applying new tools, techniques and models of New Public Management, Public Administration and Quality Governance.

Synthesizing Equity, Growth and Innovative Ideas

Many a time, we have witnessed that parochial politics, self-surviving games and backward-looking ideas have become an Achilles heel in the way of knowledge society and empowered nation. So, we will have to develop scientific temper, new ways and innovative ideas to bring attitudinal changes and transformations – in status quoists' mindset, which limit choice, freedom, entrepreneurship and institutional capacity. Concurrently, we cannot afford to become an 'amoral people' with polluted minds blocking creative responses for 'a value-based tomorrow'. So, a synthesis between technology, tradition and equitable share of water resources is recommended as a viable solution for the water-induced crises in India. The inclusive growth should, therefore, compromise between the people, the planet and the profit.

Chapter 16

Water Visions and Strategies at Global and All India Levels

Some Personal Reflections

Rakesh Hooja

A recent World Bank Document (*Water Resources and Sector Strategy: Strategic Directions for World Bank Engagement*, 2004) has pointed out that "water resources management and development are central to sustainable growth and poverty reduction". The same volume goes on to state that "most developing countries need to be active in both management and development of water resources infrastructure".

The management and development of water resources has also been considered to be vital in the *World Water Vision* prepared by the World Water Council (William J. Cosgrove and Frank R. Rijsberman, *World Water Vision: Making Water Everybody's Business, 2000*, henceforth referred to as *Water Vision*). The *Water Vision* points out that "there is a water crisis today. But, the crisis is not about having too little water to satisfy our needs. It is a crisis of managing water... badly" It adds that food production in developing countries has kept pace with population growth, with both more than doubling in the past 40 years, though the providing of six times more water now than was being done a 100 years ago has had significant impacts on people and the environment. The *Water Vision* accordingly relies on meeting half the demand for agriculture water use in 2025 by increasing water productivity with the other half of demand for water for food and rural development to be met by developing water supplies. Making water more productive ("more crop per drop") would, along with other steps like modified agronomic practices, depend upon improving irrigation water management, using more deficit, supplemental and precision irrigation, and ensuring accountability and responsiveness to users on the part of managers. Developing of water supplies shall be ensured through using a mix of large and small dams, ground water recharge, and recourse to traditional small-scale water storage techniques and rainwater harvesting (including decentralized community-based management of water resources) etc.

The *Water Vision* thus aims at empowering communities and various stakeholders and beneficiaries, producing more food and using water more productively, and managing water to conserve the quantity and quality of fresh water as also to ensure the existence of healthy catchments, encouraging more innovation in water resource management, and better governance.

The five key actions that *Water Vision* envisages to meet the above-mentioned objectives are:

- Involving all stakeholders in integrated management.
- Movement towards full cost pricing of all water services.
- Increased public funding for research and innovation.
- Increased cooperation internationally (and, may we add, even intra-nationally) in river basins.
- Massive increases in investments in water.

In a recent article in *IFPRI Forum* March, 2004 ("Five Priorities for Water Policy Reform") Mark W. Rosegrant and Claudia Ringler have pointed out that the Earth's water is under serious threat. Water bodies are being polluted, groundwater supplies are being depleted, existing supplies of water are often used wastefully and developing new sources of water is increasingly proving to be expensive. Yet growing population, expanding industries and farms, and concerned environmentalists are all demanding more usable supplies of this finite source.

Rosegrant and Ringler have suggested the following five water policy reforms to make water security a reality for everyone:

1. Establishing strong water rights and improving groundwater governance: Water rights have the potential for making water use more efficient and allocating water amongst different uses without harming the welfare of irrigators and other rural users. Secure and well defined water rights induce users to invest in water saving technology. Water rights are also part of the solution to problems of groundwater management which is a pressing issue especially in Asia. Viewing surface and groundwater as an integrated system could help make water allocation more flexible and water use more efficient.
2. Improving incentives for efficient use of agricultural water: Water pricing systems can be designed in ways that introduce incentives for efficient water use, recover operating and maintenance costs, and protect or increase farm incomes.
3. Aggressively reforming policies to improve quality: Addressing growing water quality issues will require investing in water quality monitoring, establishing measurable and feasible water quality standards for different uses, incorporating water quality standards into water rights systems, and educating developing country policy makers about water quality and environmental water uses.
4. Breeding and managing crops to conserve water: Especially breeding crops for high stress environments, like drought prone or salinized soils, can dramatically raise cereal yields without greatly increasing demand for water. Improving yields in rainfed settings also calls for changes in crop management; more water harvesting, conservation tillage, increasing nutrient management, (for paddy rice) alternative wet-dry irrigation, and the like. The greatest challenge lies in enforcing water quality standards, making polluters pay, and changing people's habit of using and discharging low quality water.
5. Stepping up research on the effects of global change on water resources: Developing countries bear the severest consequences of global change, which includes not only climate

change but also human alteration of landuse patterns, urbanization, elimination of wetlands, nutrient overloading in water systems and other bio-physical changes that could effect the ability of the global water cycle to support required food production.

A recent publication of International Food Policy Research Institute prepared by them with International Water Management Institute as part of IFRI's 2020 Vision Initiative (*World Water and Food to 2025: Dealing with Scarcity* by Mark W. Rosegrant, Ximing, Cai and Sarah A. Cline, 2002) points to the need for increasing supply for irrigation, domestic and industrial purposes through selective investments in infrastructure; for water conservation efforts and water use efficiency improvements in existing irrigation and water supply systems through water management reform, policy reform and investment in advanced technology and infrastructure; and for increasing crop productivity per unit of water and through integrated efforts in water management and agriculture research (including in rainfed agriculture). The approximate mix of water policy and management reforms and investments and institutional arrangements that are required to be made use of need to be country specific.

The World Bank Water Resources Sector Strategy referred to at the beginning of this paper also indicates that water policies and visions must be tailored according to the circumstances of each country. In this context, a look at Government of India's *Vision for Integrated Water Resources Development and Management* (Ministry of Water Resources, 2003) appears appropriate.

This Indian vision includes amongst its objectives the harnessing of the untapped water wealth of the country, addition to the existing irrigation potential created and ensuring full utilization of all created potential, additional storages, improved drainage for enhanced productivity, improved efficiency of irrigation water systems and assured serviceability of existing infrastructure, optimizing water use as per agro-climatic conditions, bringing benefits of water resources development to water scarce areas through inter-basin transfers etc. To meet the above objectives the strategies suggested include, inter-alia, the integrated management of water resources development, development of new resources, optimal utilization of developed resources, preserving the resources, stakeholders and users participation in water resource development and management, demand side management through mass awareness, dissemination of technology and research, and ensuring adequate allocation of financial outlays for the water sector.

In the backdrop of the various water visions and strategies referred to above, it is now proposed to set forth a few aspects which the author considers important elements of a water strategy for India.

The *Theme Paper on Five Decades of Water Resources Development in India* brought out by the Indian Water Resources Society in 1998 has reported assessments that by 2025 AD total fresh water requirements of India would be 1,050 b.cu.m. which would be almost at par with the country's currently estimated exploitable water resource of 1,122 b.cu.m. (690 b.cu.m. surface storage potential and 432 b.cu.m. rechargeable ground water potential). Similarly, Tata Energy Research Institute estimates quoted in newspapers in the year 2000 indicate that water demand will virtually double from 5,646 b.cu.m. in 1997 to 1,048 b.cu.m. in 2047 AD. By then the average national per capita water availability per annum would have shrunk from todays 2,200 cu.m. to 1500 cu.m. Food self-sufficiency is difficult at a run off level of less than 1,700 cu.m. per person and it appears that India will reach this stage by 2025 AD.

If per capita water availability is less than 1,700 cu.m. per annum the region is deemed to be facing water stress and if it becomes less than 1,000 cu.m. per annum then a situation of water shortage is considered to exist. A number of the country's river basins have already reached water shortage or water stressed levels. (see Chapter 1 of R.N. Athavale *Water Harvesting and Sustainable Supply in India,* and Rakesh Hooja "Administering Water in India in 2050 AD" in *Water and Energy International* Vol. 57, No. 4, October-December 2000).

A 2000 study by International Water Management Institute (IWMI) described in "Revealing the Face of Water Scarcity", *Water for Food, Water for Nature, Water for Life* Annual Report Science Programme of IWMI, Colombo indicates that water scarcity can be compounded by 'economic water scarcity', that is, by the inability of the country to mobilize funds to develop the dams, canals etc., necessary so as to tap and optimally use water resources

Thus, a major initiative required in India is to mobilize and provide the resources and to gear up the Irrigation or Water Resources Departments of the State Governments to undertake cost-effective development of water resources (storage dams, canals etc.), and to undertake steps for recharging groundwater where feasible.

This needs to be coupled with an equally strong effort at constructing water harvesting structures and adopting water harvesting practices. This would need to be undertaken jointly by the State Government's Agriculture, Watersheds, Ground Water, Panchayats, and Irrigation Departments along with Panchayati Raj Institutions and community-based organizations, user groups and even individuals. As para 12 of the National Water Policy, 2002 states "management of water resources for diverse uses' should incorporate a participatory approach by involving not only the government agencies but also the users and other stake-holders.... Water Users Associations and local bodies such as Panchayats... should particularly be involved in the operation, maintenance and management of water infrastructure facilities at appropriate levels."

Development of hitherto untapped, or partially tapped, resources needs to go hand in hand with improved water management. This would involve use of (i); an integrated, multi disciplinary, Command Area Development (CAD) Approach to managing medium and particularly large canal irrigation systems; (ii) Participatory Irrigation Management (PIM); (iii) Comprehensive Participatory Watershed Development; (iv) integrating drainage efforts into all irrigation schemes (cf. Rakesh Hooja *Management of Water for Agriculture: Irrigation, Watersheds and Drainage*, Rawat Publications, 2006); and (v) irrigation extension efforts.

Adopting an Integrated Multidisciplinary Command Area Development Approach

Western countries do not have to worry about whether a sufficiently developed physical and economic infrastructure along with provision for amenities and services exists as a given part of the environment within which an irrigation scheme can function, with the irrigation scheme itself forming the external environment for an irrigation system. Hence, for irrigation schemes such as those in India, we have to think beyond the irrigation scheme *per se* and also consider the dependency of the irrigation scheme on the larger human and natural ecosystems. Thus, irrigation schemes cannot be considered separately from area development schemes but in fact need to be considered instead as multidisciplinary, integrated and interdependent area development schemes based on irrigation and involving occupational changes, fresh settlement or resettlement of populations, infrastructural development, agriculture, irrigation, individual beneficiary schemes, and coordinated and integrated functioning of various sectoral programmes in the area.

It was with a view to ensuring coordinated, integrated, and multidisciplinary planning and implementation of irrigation schemes that India's National Water Policy of 1987 had advocated that all irrigation projects should adopt the Command Area Development (CAD) approach. Similarly, para 6.4 of the National Water Policy, 2002 states that "there should be an integrated and multidisciplinary approach to planning, formulation, clearance and implementation of projects including catchment area treatment and management, environmental and ecological aspects, the rehabilitation of affected people and *command area development*" (emphasis by the author). A CAD approach can be said to be that of multidisciplinary, interdependent and integrated area development based on irrigation and land-water management.

As Sivamohan and Scott ("Command Area Development Programme in India: A Policy Perspective" *ODI Irrigation Management Network Paper No. 20* 1992) have rightly pointed out, India's CAD concept is unique, aiming to integrate a very wide range of developmental, sectoral and functional components and sub-components through articulation of a comprehensive strategy of development which includes disparate fields previously (and, by many, even now) thought to be unrelated to irrigated agriculture.

Accordingly, Command Area Development (CAD) organizations need to be set up for all large and medium irrigation projects throughout the country. Such CAD organizations need to have the following objectives: (i) full utilization of irrigation potential,

(ii) increased productivity per unit of water and per unit of land, (iii) provision of predictable water supply to users, (iv) efficient use of water with equitable distribution at minimum administrative cost, (v) evolving appropriate cropping patterns to suit the agro-economic conditions, the available technologies and the physical and social infrastructure, (vi) conservation of soil and water resources, (vii) planned and regulated development of the command area and its inhabitants in harmony with the land, water and agro-climatic conditions, (viii) attainment of cost effectiveness and productive returns on investments for the users, (ix) protection of the developed command area to avoid environmental damage and loss of valuable command land due to waterlogging, salinity, erosion, gully formation, pest infestation, canal siltation, etc., and integrating drainage into a multidisciplinary and integrated water management strategy of preventive and curative measures to check waterlogging and salinity, (x) providing the organizational means and incentives for effective and constructive user participation in management including operation and maintenance, (xi) balancing industrial (including energy generation), domestic, livestock, drinking water and irrigation demands for water, (xii) providing basic infrastructure, improved amenities, and services, and increased employment opportunities within the command area, (xiii) integrated land and water management, and surface and sub-surface water management, through the adoption of a holistic project approach for each irrigation command area project, and (xiv) creating a coordinated and well-integrated government delivery system for the command area with appropriate emphasis on both development and management.

Propagating PIM

Participatory Irrigation Management (PIM) is more than just farmers' participation in irrigation management. It is the taking over of at least one level of irrigation channels/canals above the outlet (i.e., taking over at least a minor canal) by a Water Users Association (WUA) who take over the management of both the operation and maintenance of that canal. In addition, as part of PIM, the WUA and the irrigation agency (Irrigation/Command Area Development (CAD)/Water Resources Department) associate with each other in a joint partnership at managing the entire irrigation system. Accordingly, PIM involves a re-definition and refocusing of the government's role in irrigation and not a reduction, nor a downsizing of the irrigation agency.

The PIM approach is based upon integrated and joint management of irrigation with the community (in the form of the WUA) and government agencies working together for equitable, rational, timely and convenient allocation of water and increase of efficiencies both in canal management and in on-field water application.

The PIM approach can be applied to major, medium and minor canal irrigation systems and even to larger tank, or lift irrigation systems.

The PIM approach needs to allow for flexibility from region to region, irrigation system to irrigation system, and even WUA to WUA, and the WUAs should be free to decide about the functions that they would like to take on.

There is difference of opinion whether WUAs should only perform irrigation and canal management activities that they take over from the irrigation agency, or whether they should also perform other tasks related to irrigated agriculture.

By and large, PIM legislation in India has followed the views of those who feel that the functions of WUAs should be limited to only irrigation department related functions, viz., maintenance of the canal and its structures, distribution of water, resolving disputes related to water distribution and collection of irrigation charges.

Many experts, however, feel that WUAs can achieve institutional sustainability and both economic and institutional viability only if they undertake various irrigated agriculture related tasks including even input management and post-harvest activities, in addition to canal management.

We tend to agree with the view that WUAs should undertake a broader set of functions. What it means is that there should be two sets of tasks that WUAs should consider taking up. One, a core or compulsory group of tasks related to the management and maintenance of the canal (or lift) system which is the bare minimum that a WUA should have to take up as part of PIM and, the second, a list of additional and optional agriculture and land-water management related tasks which the WUA could adopt either at its inception or gradually add on as time passes.

The core or compulsory activities could be as follows:

- Acting as an interface between the farmers and the main system management of the irrigation project as well as other concerned government agencies (Command Area Development/Rural Development/Agriculture/Ground Water/Cooperatives Development/(Land) Revenue and Irrigation Departments).
- Water distribution, irrigation scheduling including both rotational running of canals and *warabandi*, ensuring that water reaches to all members (as also non-members) of WUAs as per their due shares.
- Operation and maintenance of the irrigation and drainage system.
- Collection (and assessment) of water charges and other user charges or special charges that the WUAs may levy.
- Resolution of local disputes amongst members, if necessary, setting up of an internal system of penalties and other incentives or disincentives.
- Conflict resolution between members and non-members.
- Drainage.
- Provision of drinking water from canals.
- Design and construction of new works (irrigation works, OFD works, drainage works) as well as rehabilitation of canals, and structures (deferred maintenance works).
- Maintenance of commercial, financial and water accounting records.
- Cooperating with other WUAs to form federations of WUAs to take over larger canal sub-systems.

The additional and optional tasks could be as follows:

- Agriculture extension and farmers training.
- Recommending of cropping patterns and package of agricultural practices suitable for the WUAs farmers.
- Helping to arrange for other inputs besides water and for credit supply to members for undertaking irrigated agriculture.

- Irrigation extension and propagation of better on-farm water application and better intra-outlet command water management.
- Encouraging and taking up of conjunctive water use, or community lift irrigation.
- Post-harvest practices (grading, packaging, storage, marketing).
- Any other task as mutually agreed upon by the members (e.g., canal side plantation, construction/maintenance of farm tracks, providing of matching funds on behalf of the command area to the District Rural Development Agency, or to Zila Parishad/Panchayat Samiti for taking up rural development work).

A WUA's functions would differ depending upon whether the WUA is for a canal system, or for a largish tank scheme, or for a community lift irrigation system.

Comprehensive Participatory Watershed Development

Watershed development has long been considered the main catalyst for all development activities in rainfed areas. In a manner, the watershed development is to dry rainfed areas what the Command Area Development is to canal irrigated areas except that since in canal irrigated command areas the role of government agencies is greater, a major governmental interdisciplinary CAD organization is required for each large, medium or major projects. For watershed development the role of the village level institutions or community-based organizations (CBOs) like watershed committees, watershed associations, user groups, self-help groups comprising agriculturists, livestock breeders, landless labourers, local developers of pastures and grazing grounds etc., is greater. These CBOs interact with project implementation agencies (PIAs) who may be panchayat bodies, government agencies, or NGOs and carry out the field level activities with coordination and supervision being provided from district and state levels.

Some of the objectives of watershed development as per Government of India guidelines of 2001 and 2003 include the mitigating of the adverse effects of drought and desertification, employing simple, easy and affordable technical solutions and institutional arrangements; encouraging the village community towards sustainable community action for operation and maintenance of assets and development of the natural resource potential of the watershed; promoting overall economic development and improving the socio-economic condition of the resource poor and disadvantaged sections, employment generation, poverty alleviation, community empowerment and development of human and other economic resources in rural areas, restoring the ecological balance by harnessing, conserving and developing natural resources like land, water, vegetative cover, developing wastelands, degraded lands, drought prone and desert areas keeping in view the capability of the land and local needs and conditions, harnessing every drop of rainwater for irrigation, horticulture, floriculture, pasture development/fisheries etc. to create sustainable sources for the village community and provide drinking water supplies etc.

The above set of objectives may be compared with those worked out by the author in a list of objectives initially developed in 1979-80 (and published in Rakesh Hooja, "People's Participation in Comprehensive Watershed Area Development" in Hoshiar, Singh (ed.) *Rural Development in India*, Printwell 1985) and then slightly revised for the author's contributions to *Social and Institutional Issues in Watershed Management in India*, (OIKOS Delhi and IIRR Philippines, 2000). The list reads as follows: (i) conservation of moisture in rainfed areas for optimal production, (ii) to check soil erosion, ensure soil and water conservation, (iii) to control problems of drainage, salinity, alkalinity, (iv) flood prevention, (v) to check siltation in reservoirs, (vi) collection of surplus run off in farm ponds and its recycling for crop use, (vii) to recharge groundwater or increase the water table in wells, (viii) collection of surplus run off for meeting the drinking water requirement of livestock and human population in desert areas or for partially irrigating a small patch of land, (ix) to improve the main and on-farm irrigation systems for increased productivity and increased area under irrigation, (x) to achieve optimum productivity per unit land and per unit water, (xi) the balancing of non-agricultural (industrial, domestic and human settlement) uses of land and water with agriculture, animal husbandry and allied uses, and (xii) generation of income and employment in harmony with the land and agro-climatic conditions of each region or sub-region.

Watershed development as being advocated above encompasses watershed management and para 3.4 of National Water Policy, 2002 has recommended that watershed management should be promoted.

Integrating Drainage Efforts into all irrigation Schemes

The time has now come for drainage to be brought to the centre stage in all irrigation efforts, due to (a) farmers having invested significantly in irrigated agriculture and thus not wanting to loose productive lands or to endure lower yields and thus lower incomes because of water logging/salinity etc., (b) many of the irrigation systems developed in the last 50 years, or even earlier having started facing problems of water logging etc., (c) the fact that at the time of irrigation development, several decades back, drainage industry was in a nascent stage but now it has evolved sufficiently to be able to offer commercially viable solutions for the field level problems, and (d) conditions of water stress prevail now in many areas inducing farmers to try and obtain maximum productivity per quantity of water utilized as also maximum productivity from their developed irrigated lands.

Not only do drainage and sub-surface drainage works need to be part of every irrigation project, but many preventive and curative measures which involve better on farm water application, or efficient management of canals, are required to be introduced to prevent waterlogging and to supplement the engineering efforts of drainage. Thus drainage has become a vital part of water management activities.

Irrigation Extension

Before ending it needs to be pointed out that many of the measures advocated in this paper would benefit from the introduction of what this author likes to refer to as "irrigation extension". Just as we have for many years now had the concept of agriculture extension (Rakesh Hooja "Transforming Agriculture Extension and Research to Meet Contemporary Needs" in V.S. Mahajan (ed.) *Political Economy of Reforms and Liberalization*, (Vol. 2), Deep & Deep Publications, New Delhi 1997) to help in the spread of selected traditional agricultural practices from one region to the another, or in the transfer of new technologies related to agronomy, plant protection etc. from research laboratories to the farmers, similarly, the need has arisen to launch a significant irrigation extension effort related to improved water management and water application practices in canal irrigated, groundwater irrigated as well as rainfed areas.

And, finally, it is strongly felt that, while implementing the water vision and strategies considered appropriate for the country and for its various agro-climatic regions, and while undertaking water resources development and utilization planning and implementation, there is also a need to see irrigated areas and rainfed or dry farming areas (both of which cut across village panchayat or district boundaries) as part of one interrelated continuum.

Chapter 17

Challenges and Priorities for Water and Sanitation Sector Reforms for Urban Areas

K.K. Gadeock

As per the *World Bank Report* (2005) "India's Water Economy: Bracing for the Turbulent Future", India faces a turbulent future unless water management practices are changed soon. India will face a severe water crisis within the next two decades. India will neither have cash to build new infrastructure nor the water needed by its growing economy and rising population. The report further mentions that the infrastructure is crumbling and there is enormous backlog of investment due to shortfalls in financing.

The water and sanitation trends in India testify to the observations contained in the report. Increased investment has not led to sustained improvement in services. Water supply coverage has been increasing but the service is unreliable and the quality is poor. Sanitation coverage may be increasing but most sewerage is not properly collected or treated. Investment in solid waste management have not led to satisfactory service provision, safe disposal is virtually absent.

The sustainability challenge in water and sewerage sector (WSS) is due to deteriorating assets, declining productivity, increasing costs, low tariff, declining revenues, unpaid bills, inability to access financing and declining credit rating and investment.

There have been repeated investments in infrastructure. Increased planned investment has not led to sustained improvement in service delivery and the availability of water is in decline.

Table 17.1

Hours of Water Supply Per Day in Some Cities

Delhi	4 hours
Jodhpur	2.5 hours
Jaipur	3 hours
Ahmedabad	2 hours
Bikaner	1.5 hours
Bangalore	2.5 hours
Udaipur	2.5 hours
Chennai	1.5 hours

Table 17.2

Position of Water and Sanitation Sector in Urban India (2001)

Water Services	
Access to safe drinking sources	82%
Access to tapped drinking water	65%
Access to tapped drinking water within premises	42%
Average per capita supply	54 to 150 lpcd
Sanitation	
Access to toilets	63%
Access to toilets within premises	33%
Connections to sewerage	28%

Further, unaccounted for water (UFW) accounts for 25 per cent to 50 per cent, which is the reason for high operating cost. Transmission and distribution are in poor conditions and are poorly maintained.

Table 17.3

Position of Cost Recovery of Select Cities

	Revenue as a percentage of O&M Cost	
City	*March-1998*	*1999-2000*
Kolkata	14	
Nagpur	48	
Pune	49	
Ludhiana	40	
Lucknow	50	
Kanpur	78	
Delhi	26	43
Hyderabad	230	
Mumbai	268	
Chennai	184	

Source: India Water Supply and Sanitation Assessment 2002,WHO-UNICEF.

Around 30-60 per cent of the urban population living in slums has limited or no access to sanitation facilities. Access to sewerage facility is very low and open defecation is prevalent. Only 700 out of 5,161 cities treat their effluent before disposal in the environment.

In case of solid waste management (SWM), organized collection has improved, solid waste transfer in closed vehicles has slightly improved, and street sweeping has substantially improved but sanitary landfill for disposals are negligible.

The Central Public Health and Environmental Engineering Organization (CPHEEO) norms for water supply are as under:

- 150 lpcd for mega cities
- Between 135-150 lpcd for million plus cities
- 135 lpcd for Cities below million population
- 70 lpcd for Guwahati (cities below million population)
- Unaccounted for water: 15 per cent across all categories

The water sector reforms aim at 100 per cent area coverage, 100 per cent population coverage, 24 hours supply of water, achieving unaccounted for water as per norms and 100 per cent cost recovery of O&M. We are far short of this!

The existing institutional models of service delivery for water and sanitation sector in this county are:

Figure 17.1

Existing Models of Service Delivery

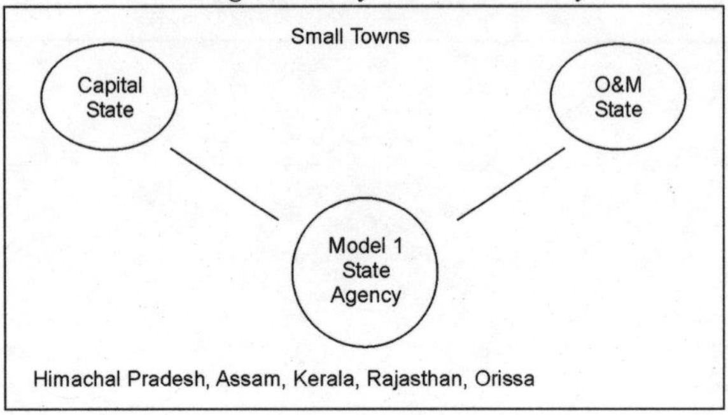

Figure 17.2

Existing Models of Delivery Small Towns

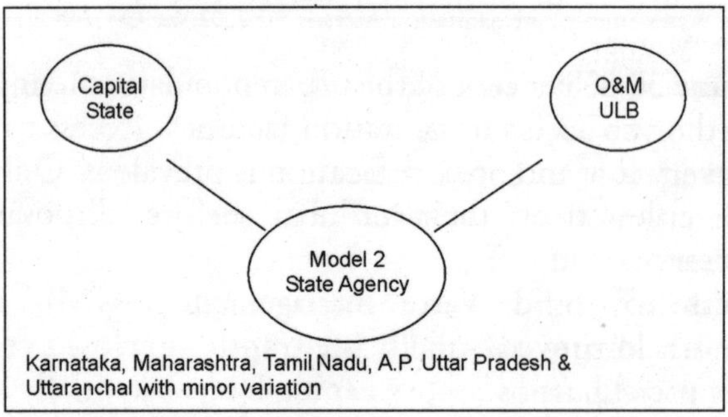

Figure 17.3

Existing Models of Delivery for Metros and Large Cities

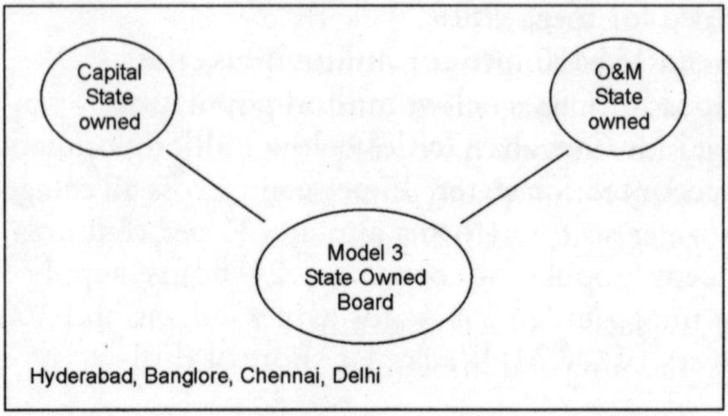

Figure 17.4
Existing Models of Delivery for Metros and Large Cities

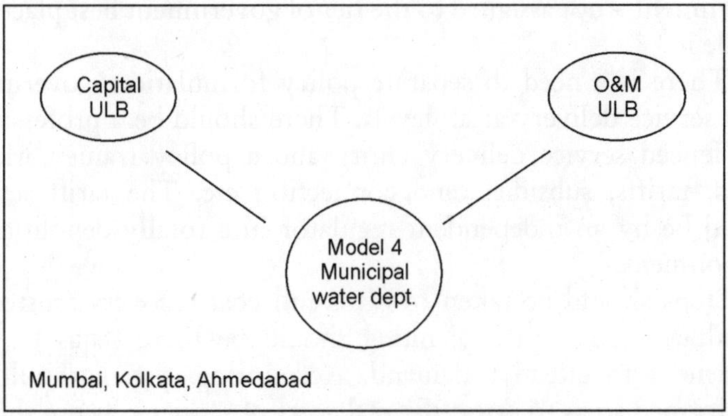

Further, sewerage network coverage in most cities is low and areas covered by sewerage networks have poor maintenance. Treatment capacity in plants is lower than the generated waste leading to sewerage discharge in rivers.

The service delivery is supply driven and the public agency delivering the service is far removed from the users and does not have the mandate or incentive to be responsible to their needs. The service delivery model is not sustainable.

Poor service standards and financial crisis in WSS sector co-exist with high burden on the state exchequer – government funds are subsidizing the inefficiencies of utilities rather than serving any social purpose.

The answer perhaps lies in the lack of accountability and the incentive inherent in the institutional arrangements through which WSS services are delivered – monolithic centralized WSS institutional structure (PHED) with project linked budgetary support for asset creation with limited attention to efficient and sustainable asset management and operations. Repeated huge government subventions have supported the inefficiency of the service providers rather than the targeted support to the poor who are largely unserved and have to cope by incurring significant expenses.

There is lack of accountability because people are not involved in planning and implementation of the projects, fortunes of service providers are not linked to the customer satisfaction. Further, WSS responsibility has not been devolved to local government – not assigned to the tier of government best placed to handle it.

There is a need to separate policy formulation, governance from service delivery at all levels. There should be a professional ring fenced service delivery entity and a policy framework on access, tariffs, subsidies, and connection etc. The tariff setting should be by an independent regulator in a totally depoliticized environment.

Steps should be taken towards full cost recovery, customer consultation in service planning should be there so as to link investment to effective demand. Accounting system should be transformed so as to assess the real costs of services. There should be a system of indexation, which would enable revenue to keep up with the rising cost.

It is essential to recognize and reduce inefficiency related to excess manpower, poor collection of revenues and higher levels of leakages in the system. Many utilities can actually recover the costs if theft, billing and collection are at acceptable level.

The World Bank, in its report, *India Water Supply and Sanitation: Bridging the Gap between Infrastructure and Service* January, 2006, has given the following five suggestions:

1. Improve accountability of institutions to improve reliability of service:
   (i) Clarify the role of actors: There is an unhealthy overlap between policymaking, regulation, and financing ownership of infrastructure and operation of service within state agencies. The way out is unbundling of these services, establishing clear terms of references and building accountability mechanism.
   (ii) Devolve full responsibility for providing the service to the most appropriate level of government, i.e., ULBs.
   (iii) 'Ring fence' WSS operations.
   (iv) Transform beneficiaries into paying customers.
   (v) Look for approaches that strengthen performance incentives, including an enhanced private sector role.
2. Improve the financial sustainability of the service provider:
   (i) Price the WSS service according to sound principles.

(ii) Finance the transition towards full cost recovery in a transparent manner.
(iii) Create financial incentives to reduce O&M costs, capital and financing costs.
(iv) Expand the role of programmes sponsored by GOI to support reform.
(v) Encourage the use of other sources of financing to promote reform.

3. Improve the environmental sustainability of service:
(i) Price water according to sound economic principles: This usually means using long run marginal cost (LRMC) pricing. Estimates of LRMC should include the cost of addressing environmental externalities, such as the proper collection and disposal of wastewater.
(ii) Strengthen water rights and develop water markets.
(iii) Limit depletion of ground water.
(iv) Plan infrastructure to achieve realistic environmental objectives.,
(v) Protect water quality
(vi) Support behavioural changes towards better sanitation practices.

4. Improve the affordability of services:
(i) Reduce costs by adapting design criteria and technology.
(ii) Reduce costs by adapting operation and maintenance arrangements.
(iii) Implement a cost recovery strategy adapted to the poor.

5. Build the capacity of the WSS sector:
(i) Build an identity for urban WSS industry.
(ii) Adapt training programmes to the new needs of the urban WSS sector.
(iii) Develop special information programmes for key stakeholders.

Water has not been transferred to the municipalities in Rajasthan as required under 12th Schedule of 74th (Constitutional Amendment Act and PHED continues to provide service in the state. The financial status of PHED, Jaipur is as under:

Financial Status of PHED

1999-2000	2000-01	2001-02	2002-03
Loss (-) 35.4 crores	(-) 43.4 crores	(-) 44.8 crores	(-) 50.5 crores

The financial problem mentioned above is caused by:

- High NRW
- Low tariff levels insufficient to recover direct costs
- Low operational efficiencies

The amount of NRW (non-revenue water) in Jaipur is calculated at around 50 per cent of the total amount of water produced. The current tariff table has not been revised since 1998. If the tariff table is revised according to a 6 per cent inflation rate, the current tariff level can be estimated as described below:

	Present tariff levels (2004)	Estimated tariff revised with inflation (Rs per cubic metre)
Average tariff for domestic customer	2.8	4.0
Average tariff for commercial customer	7.2	10.2

On an average, domestic consumers in Jaipur pay Rs 150-200 per month for electricity consumption and a similar amount for use of telephone. Compared to this, the average bill of water per month per household is only Rs 58.

Table 17.4

Tariff Comparison Across Key Cities

Bangalore		*Hyderabad*		*Guwahati*		*Chennai*		*Jaipur*	
Consumption	*Rate*	*Consumption*	*Rate*	*Consumption*	*Rate*	*Consumption*	*Rate*	*Consumption*	*Rate*
Minimum	*90*	*Minimum*	*90*	*Minimum*	*110*	*Minimum*	*50*	*Minimum*	*16*
<15 m3	6	<30 m3	5	<30 m3	7	<10 m3	2.5	<15 m3	1.25
15-25 m3	8	200-300 m3	10	30-50 m3	9	10-15 m3	10.0	15-40 m3	2.4
25-50 m3	12	>200 m3	25	>50 m3	11	15-25 m3	15.0	>40 m3	3.2
50-75 m3	30					>25 m3	25.0		
>75 m3	36								

The cost recovery comparison with other cities in India reveals that some cities are fully recovering the direct operation and maintenance costs, whereas Jaipur is recovering only around 34 per cent:

	Jaipur	*Bangalore*	*Chennai*	*Hyderabad*
Cost recovery	34%	101%	121%	100%

Table 17.5

Cost Subsidy Indicators for the Financial Year 2003-04

Classification	*% Consumption*	*Revenues (Rs millions)*	*% Billing*
Domestic	81%	134	63%
Commercial	16%	68	32%
Industrial	1%	7	3%
Flat rate	2%	3	1%
Total	100%	212	100%

Commercial and industrial consumers account for 17 per cent of the water consumption in Jaipur and contribute around 85 per cent of the total revenue of PHED, Jaipur. Domestic consumers, in turn, consume more than 80 per cent of water, which contribute only 63 per cent of the total revenue. This indicates that commercial and industrial sectors are already being penalized for consuming a scarce resource.

Further, it has been found that effective rate for lower consumption levels are higher under domestic category. Effective rates mean the actual unit rates calculated by dividing the amount of water charges by the amount of water consumed. A poor household consuming 10 cubic metre pays an effective rate of Rs 1.68 per cubic metre, whereas a household consuming 21 cubic metre pays an effective rate of Rs 1.58 per cubic metre. In addition, the effective costs per cubic metre in higher consumption ranges are lower than those in the lower consumption ranges. This means that in terms of resource conservation, the current tariff system does not provide much incentive to the consumers to consume less water, particularly scarce resource in the context of Rajasthan. So, there is an urgent need to change the existing tariff structure.

In "Aapni Yojna" being implemented in rural areas of Churu, Hanumangarh and Jhunjhunu districts of Rajasthan, the beneficiaries are paying the tariff for public stand post, which is almost three times the tariff prevalent in Jaipur for water supply within the premises. Many studies have shown that willingness to pay is there, but willingness to charge is not there.

SAPI (Special Assistance for Project Implementation) for Bisalpur-Jaipur Water Supply Project, in its report issued October 2004, had considered the following options for Jaipur water supply:

Table 17.6

Possible Options for Jaipur Water Supply

Activities	*Option 1*	*Option 2*	*Option 3*
Policy planning	PHED	PHED	PHED
Regulation	PHED	Independent	Independent
Development planning & construction (including new tubewells)	PHED	PHED	PHED
Bulk Transmission from Bisalpur	PHED	PHED	PHED
City distribution (including existing tubewells	JMC	JMC- contracted to PHED, Jaipur	Autonomous Corporation
Customer services	JMC (outsourced)	JMC-contracted to PHED, Jaipur (outsourced)	Autonomous Corporation (outsourced)
Sewerage treatment	JMC	JMC-contracted to PHED, Jaipur	Autonomous Corporation
Sewerage O&M of sewer lines	JMC	JMC	Autonomous Corporation

SAPI has recommended the option 3 for Jaipur water supply. This option recommends setting up of autonomous corporation for Jaipur water supply for city distribution, customer services, sewerage treatment and O&M for sewer lines while retaining policy planning, development planning and construction (including new tubewells), bulk transmission from Bisalpur with PHED. It has also recommended setting up of an independent regulator.

The recommendations are under active consideration of the Government of Rajasthan. The recommendations of SAPI are relevant in the present context as the countries that have reformed have done so by following the corporatization model, effective regulatory agency, incentive based contracts and private sector participation. The sooner the Rajasthan government brings about institutional and organizational reforms, the better it would be. There is an urgent need for a regulator also in the urban sector especially for water and sanitation sector.

Reform in water sector in Jaipur will usher in reform in the state as a whole. Jaipur has to provide lead in this field. For water and sanitation as a whole, apart from the institutional and organizational reforms, the government will have to engage itself in critical areas such as ground water management, establishment and management of water entitlement and financing of public goods such as flood control and wastewater treatment.

There is a need to bring abstraction in ground water in line with recharge. The end of era of massive expansion in ground water use is going to demand greater reliance on surface water supply system. There is an urgent need for dramatic transformation in the way in which public water services are provided to farmers, households and industries – financial sustainability, accountability, competition, regulation and entry of alternatives to government provisions, including cooperatives and private sector. Levy of an appropriately designed system of access and usage charges and implementing it strictly should be a national priority. Model bills for ground water regulation are pending with policy makers in the state and needs to be activated immediately.

Further, it is the right time to make the service sustainable. People are already paying a lot for water, which can be redirected to the utility as services improve. It is high time that governments at all levels realized that their role of being a key player in the water in the last six decades has been a failure. They have always looked upon themselves as builders of water assets and never as providers of water service to the users.

Funds are available for improvement under JLNURM. Domestic and international financial institutions are ready to finance bankable projects. Most cities have enough water available – allocation and conservation should be the priority. Talent and skills are available. Indian private companies are prepared to engage in the sector. Even civil society is also prepared to engage.

■ References

1. *India's Water Economy: Bracing for a Turbulent Future*, World Bank (2005).
2. *India Water Supply & Sanitation: Bridging the Gap between Infrastructure & Service*, World Bank Report (2006).
3. SAPI (Special Assistance for Project Implementation) for Bisalpur-Jaipur Water Supply, October 2004.
4. "Expanding Access to Water Supply and Wastewater services to the Urban Poor: International Approaches & Challenges" by Jim Woodcock, Water & Sanitation Consultant, World Bank, Jakarta.
5. "Best Practices in Action: International Experience: Governance, Policy, Regulation and Effective Management" by Allen Eisendrath.
6. *Characteristics of Well Performing Public Water Utilities:* The World Bank, February 2006.
7. Urban Water Sector in South Asia: Benchmarking Performance, Water & Sanitation Programme, May 2006.
8. "Is a serious water crisis inevitable?" by Amrit Pandurangi.
9. "Profile of Current Policy and Regulatory Environment in India" by N. Bhattacharjee, Office of Economic Growth, USAID, INDIA.

Contributors

RAKESH HOOJA, Director, HCM Rajasthan State Institute of Public Administration, Jaipur.

RAMESH K. ARORA, Visiting Professor, HCM Rajasthan State Institute of Public Administration, Jaipur.

K.K. PARNAMI, Publication Officer, HCM Rajasthan State Institute of Public Administration, Jaipur.

YOGINDER K. ALAGH, Professor and Vice Chairman, Sardar Patel Institute of Economic and Social Research, Ahmedabad.

S.A. KULKARNI, Director, International Commission on Irrigation and Drainage (ICID), New Delhi.

BRIJ GOPAL, School of Environmental Studies, Jawaharlal Nehru University, New Delhi.

G.S. NARWANI, Former Special Secretary, Rural Development, Government of Rajasthan, Jaipur.

M.V.S. PRASAD, Assistant Professor, College of Management Studies, GITAM, Vishakhapatnam.

A.S. KAPOOR, Former Chairman, Indira Gandhi Nahar Board, Jaipur.

KALYAN PAUL, C/o Anita Paul, Director (Community Initiatives), Pan Himalayan Grassroots Development Foundation, Post Bag 3, Ranikhet, Almora, Uttranchal.

C. SEKAR, Associate Professor (Agricultural Economics), PHCT, Agriculture Engineering College and Research Institute, Coimbatore.

S. SENTHILNATHAN, Ph.D. Scholar (Agricultural Economics), Tamil Nadu Agricultural University, Coimbatore.

RAKSHAT HOOJA, Programme Coordinator, Minnesota Studies in International Development–India Programme, Jaipur.

D.K. VERMA, Associate Professor and Head, Division of SC, ST and OBC Development, Dr Babasaheb Ambedkar National Institute of Social Science, Mhow.

A. SOHRUT, Research Associate, Dr Babasaheb Ambedkar National Institute of Social Science, Mhow.

RAJKUMAR SIWACH, Faculty, Department of Public Administration, Choudhary Devi Lal University, Sirsa (Haryana).

K.K. GADEOCK, HUDCO Professor of Urban Development, HCM Rajasthan State Institute of Public Administration, Jaipur.